Japanese Robot Culture

Yuji Sone

Japanese Robot Culture

Performance, Imagination, and Modernity

Yuji Sone
Macquarie University
Sydney, Australia

ISBN 978-1-137-53216-9   ISBN 978-1-137-52527-7   (eBook)
DOI 10.1057/978-1-137-52527-7

Library of Congress Control Number: 2016958023

Cover illustration: © mauritius images GmbH / Alamy Stock Photo

Printed on acid-free paper

This Palgrave Macmillan imprint is published by Springer Nature
The registered company is Nature America Inc.
The registered company address is: 1 New York Plaza, New York, NY 10004, U.S.A.

For Meredith Morse

ACKNOWLEDGEMENTS

This book would not have been possible without the help of many people. I am extremely grateful for the support I received when I was grappling with the question of robot performance from an interdisciplinary perspective. This book project is indebted, first of all, to senior colleagues who encouraged me to pursue it. I am appreciative of institutional support that enabled me to get this project underway and offered me the time to complete it. My special thanks go to Joseph Pugliese, my colleague at the department of Media, Music, Communication, and Cultural Studies at Macquarie University, who inspired me to turn my research interest on Japanese culture and the figure of the robot into a book, offering insightful comments on my material in the book's early stages. I am grateful to have received initial funding support from the university through the new staff grant at the time that I started my position in cultural studies (when the department was configured differently). I am obliged to Ann Cranny-Francis and Nicole Anderson for their support for my application. I received a grant in the development phase of this project from the Japan Foundation, through a fellowship programme that funded travel to Japan. I would like to express my gratitude to Peter Eckersall for his support for the programme and for his expert feedback on my research, which constructively shaped my outlook as I was writing. I am extremely grateful for Norie Neumark's assistance in the preparation of my book proposal and for providing me with an opportunity to present my research. Norie has supported my work since my doctoral project.

This book has had a long gestation period: its topics, concerns, and focus arose from several years of what might be called non-linear research on performance, technology, and Japan, prior to and alongside my more focused efforts. I was fortunate to receive helpful advice from, and benefit from the goodwill of, a variety of individuals across institutions. I would like to express my appreciation to Satoshi Iwaki, Hiroshima City University, who invited me to give a talk to his fellow robotics colleagues in Japan, and Hayato Kosuge, at Keio University, who gave me an opportunity to present my research to Japanese theatre studies colleagues. I am thankful for support on many occasions from the Japan Foundation, Sydney, especially that of Wakao Koike, who backed this project in its early days, and Elicia O'Reilly, who included me in a public talk series at the Art Gallery of New South Wales, Sydney, on Japanese art and performance. I thank Kathy Cleland, Christine de Matos, Julie-Anne Long, and Nikki Sullivan for providing me with opportunities to give talks about my project at their organised events. I would also like to express my gratitude for the support I received from Mark Evans, Helena Grehan, Julian Knowles, Catharine Lumby, Ian Maxwell, Katherine Millard, Edward Scheer, and Cathryn Vasseleu.

This book includes some material, in revised form, that was originally published in the following articles and chapters. I would like to thank the editors (and anonymous reviewers) who provided me with stimulating and instructive feedback. These publications are as follows:

2008. 'Realism of the Unreal: The Japanese Robot and the Performance of Representation.' *Visual Communication* 7(3): 345–62.

2010. 'More than objects: Robot performance in Japan's Bacarobo Theatre.' *Studies in Theatre and Performance* 30(3): 341–53.

2012. 'Between Machines and Humans: Reflexive Anthropomorphism in Japanese Robot Competitions.' *About Performance*, no. 11: 63–81.

2012. 'Double Acts: Human-Robot Performance in Japan's Bacarobo Theatre.' In *A World of Popular Entertainments: An Edited Volume of Critical Essays*, edited by Gillian Arrighi and Victor Emeljanow, 40–51. Newcastle upon Tyne: Cambridge Scholars Publishing.

2013. 'Robot Double: Hiroshi Ishiguro's Reflexive Machines.' In *Handbook of Research on Technoself Identity in a Technological Society*, edited by Rocci Luppicini, 680–702. Hershey, Pa.: IGI Global.

2014. 'Canted Desire: Otaku Performance in Japanese Popular Culture.' *Cultural Studies Review* 20(2): 196–222. DOI: http://dx.doi.org/10.5130/csr.v20i2.3700.

2014. 'Consumable Voice: Japanese Virtual Diva Hatsune Miku.' In *Voice/Presence/Absence*, edited by Malcolm Angelucci and Chris Caines, n. pag. Sydney, Australia: UTS ePress. DOI: http://dx.doi.org/10.5130/978-1-86365-431-9.

2014. 'Imaginary Warriors: Fighting Robots in Japanese Popular Entertainment Performance.' *Australasian Drama Studies*, no. 65: 255–71.

I would like to thank my family: my late father, Kenji, my mother, Yoshi, and my brothers, Kenya and especially Toyoji who went out of his way to offer assistance and support.

This book is dedicated to my partner Meredith Morse. There is much to thank her for: our lively discussions of ideas and for always being the first reader of my work.

Contents

LIST OF FIGURES

CHAPTER 1

Introduction: The Japanese Robot and Performance

In this book, I argue that the widely discussed Japanese affinity for the robot is the outcome of a complex loop of representation and expectation that is also an ideological iteration within Japan's continuing struggle with modernity. It is theatrical, I suggest, because the popular view of the robot in Japan is expressed in terms of the operations of theatre, through concepts such as representation, actor, audience, and setting or *mise en scène*. It is also performative in the sense that the mainstream notion of the robot in Japan is imaginatively maintained through socially enacted reinterpretations and recreations.

This book aims to provide an alternative to a stereotypical view of affinities between the Japanese and robots, that is, an approach that differs from the established discussion of Japanese next-generation robots, which presumes a static idea of Japan or Japanese culture. What the book seeks to highlight are the performative mechanisms of particular *mises en scène*. Through a performance studies approach that sees social expression as performatively producing identity and culture, the book shines light on the processes of re-signification and reinvestment concerning the idea of the robot in Japan.

Japan has been embracing robotic technologies in its manufacturing sector since the late 1960s, which contributed to the country's economic success. When Japan's economy was booming in the 1980s, factories rapidly pursued automatisation and robotisation. According to writer and translator Frederik L. Schodt, the view outside Japan of 'Japan as a Robot Kingdom' was seen within Japan with 'intense pride in manufacturing and

© The Author(s) 2017
Y. Sone, *Japanese Robot Culture*,
DOI 10.1057/978-1-137-52527-7_1

technology' in that period (1988, 15). A decade or two later, in the new millennium, the Japanese government and the robotics industry turned to addressing a labour shortage due to the country's shrinking and rapidly ageing population. To that end, the prototypes of new robots, designed to work in places other than factories, such as hospitals, the office, and the home, were developed and showcased in large corporate and government exhibitions in the 2000s. Development of these robots has continued to the present, supported by government initiatives.

These robotic machines appear as anthropomorphic and zoomorphic figures.[1] The human-shaped robots are called 'humanoids' or 'androids'.[2] While humanoids have metallic surfaces and the basic form of human bodies, androids are not only human-like in body shape but are equipped with skin-like surfaces and often have detailed facial features. These new types of robots are variously called 'social robot', 'service robot', 'communication robot', or 'partner robot', and they feature in a new field within robot engineering often called 'social robotics'.[3] Japan's focus on the development of these next-generation robots is widely acknowledged in the English-speaking world, in both popular science books (Wood 2003; Hornyak 2006; Levy 2007; Benford and Malartre 2008; Meadows 2011) and in science and engineering contexts (Menzel and D'Aluisio 2000; Perkowitz 2004; Bekey et al. 2008; Bar-Cohen and Hanson 2009; Guillot and Meyer 2010; Kaplan 2011; Ford 2015). Social robotics is also critically discussed in the humanities and social sciences (Geraci 2010; Wallach and Allen 2009; Lin et al. 2012; Turkle 2011; Šabanović 2014; Sandry 2015) as well as in a notable series of essays by Japanese studies scholar Jennifer Robertson (2007, 2010, 2014). Recent book-length studies on Japanese manga/anime (LaMarre 2009; Brown 2010; Condry 2013) and on Japanese toys and gadgets (Allison 2006) include a discussion of the cultural significance of humanoid and zoomorphic robot themes.[4]

The development of the social robot is very much the product of an iterative relation with the social imaginary, itself the result of the history of Japanese culture after World War II. The wide variety of media and cultural products that has arisen, including manga and anime, actual robotic machines, and government and corporate staging of robot events, have complexly shaped contemporary Japanese robot culture. Are there connections between the cultural conception of the robot in Japan and the technology? This book focuses on the 'performing' robot in both its actuality and as a potent figure of the social imaginary. I employ the critical lenses of theatre and performance studies, not usually deployed in discussions of

Japanese robot culture in all of its cultural and social expression, to reveal the deep-seated assumptions and ideological perspectives that have shaped and continue to shape the popular reception and conceptualisation of the robot. I argue that the social meaning of the robot for the viewing public is, in essence, activated through carefully structured, theatrical *mises en scène* (staging) that always surround the appearance of robots in Japanese public (and private) life. I use 'performing robot' to mean the robot—usually an anthropomorphic or zoomorphic robot—as an agent in these staged situations that enacts and reinforces socially established ideas (and furthers official agendas). The performing robot is both the actuality and the imagined figure, completing a circuit of meaning concerning Japanese progress and futurity that is made fun and entertaining for the ready consumption of the viewing public.

In recent years, along with the innovative uses of digital media technologies in theatre, there have been numerous book-length publications in theatre and performance studies that look into art practices that deploy robotic technologies (Dery 1996; Smith 2005; Berghaus 2005; Dixon 2007; Causey 2006; Giannachi 2004, 2007; Broadhurst 2009; Birringer 2008; Salter 2010; Parker-Starbuck 2011; Klich and Scheer 2012). As the human body is often the focus of theatre and performance studies analyses, scholars discuss the intertwinement of the body with technologies. Scholars have used terms such as 'cyborg', 'posthuman', 'virtuality', 'argumentation', or 'interactivity' to explain their investigations. Of relevance to the current study, Dixon (2007) and Salter (2010) include a chapter on 'performing machines' in their discussions. They look at artists and art groups that produce performance works that use mechanical or anthropomorphic robots as 'performers'.[5] These robotic performances function on the basis of the tension between 'the *mimetic* (imitative of human behavior in appearance) and the *machinic* (electromechanical behavior that, though animate, is not anthropomorphic)' (Salter 2010, 277, original emphasis). In a way, these artistic performance makers exploit the robot's ambiguity as a kind of performer and the audience's ambivalent feelings toward them. I am primarily concerned with the Japanese next-generation anthropomorphic and zoomorphic robots that are mimetic, per Salter's explanation.

Scholars are exploring this very ambiguity, asking in what ways robots can be performers. For example, with the term 'camp', Dixon argues that '[r]obotic movement mimics and exaggerates but never achieves the human, just as camp movement mimics and exaggerates, but never achieves womanhood' (2007, 274) and that camp is an essential factor in anthropomorphic and zoomorphic robot performance. Performance theorist Jennifer Parker-Starbuck uses the term 'abject', pointing to 'a productive tension between the subject and the object, serving to disrupt the fixity around these terms', when examining a robotic transformation of a human character in Richard Maxwell's theatre piece titled *Joe* (2011, 54). Performance theorist Philip Auslander, on the other hand, discusses examples of performance art in which there is no difference in overall artistic intention whether tasks are carried out by a human performer or robot performer, and where the actions of a human or a robot can be regarded equally as art performances (2006, 98). Auslander highlights '[cultural] anxiety in [the] emphasis on liveness and implicit resistance to the concept of machine performers' in traditional understandings of performance (2006, 90). Under the collaborative project banner of 'New Media Dramaturgy', theorists Peter Eckersall (2015), Edward Scheer (2015), and Helena Grehan (2015) are working to theorise the dramaturgical aspects of artist Kris Verdonck's robotic works in terms of robot agency and new forms of spectatorship. Celebrated playwright and director Oriza Hirata and renowned roboticist Hiroshi Ishiguro have collaborated on theatre projects since 2008 in which Ishiguro's humanoids and androids are used in Hirata's theatre works. These works have provided concrete examples for examinations by theatre studies scholars (Hibino 2012; D'Cruz 2014; Poulton 2014; Parker-Starbuck 2015; Eckersall 2015).[6] (I will discuss the Hirata and Ishiguro collaboration in Chap. 4.)

Though theatre and performance scholars have looked at 'high art' forms, so far, no studies within the theatre and performance literature treat popular forms, such as robot competitions, public installations, corporate expos, and domestic interaction with robots. I discuss these engagements with the Japanese public in this study. Overall, the Japanese audiences in popular contexts are not expecting the appearance or performance of the robots to be challenging, subversive, or ironic, and that is certainly not what the presenters intend either. The audience is instead appreciating the robots as performers, facilitated through spectacular or otherwise convincing stagecraft. The robots appear and are understood as the embodied presence of established ideas of the robot from familiar sources, such as manga/anime. Throughout this book, I explore this sense of robots as simultaneously real and representation, which is essential to an understanding of the Japanese perception of these performances presentations.

The next two sections of this chapter provide some background to the themes I address in this book. The first section that follows treats the robot in terms of Japanese history, and a Japanese approach to modernity and technology. In the second section, I return to key concepts in theatre and performance studies, inflecting them with particularly Japanese understandings of the way that objects can perform and of the role of the spectator. In the last section of this chapter, I provide an outline of the remaining chapters in this book.

JAPANESE CULTURE AND THE ROBOT

Defining the Robot

The interdisciplinarity of academic fields that deal with topics to do with robots and robotics reflects the ambiguity of the term robot. The word robot comes from Czech playwright Karel Čapek's play, *R.U.R.* (*Rossum's Universal Robots*), published in 1920. Čapek had created the term from *robota*, a Czech word for 'labour'. In its current usage, however, to refer to a robot has several meanings. For example, the first three definitions of the robot in the Oxford English Dictionary (OED (Online) 2000) are: 'An intelligent artificial being [in science fiction] typically made of metal and resembling in some way a human or other animal'; 'A person who acts mechanically or without emotion'[7]; 'A machine capable of automatically carrying out a complex series of movements, *esp.* one which is programmable'. The third meaning in OED, designating the term as it may relate to the robot industry, needs to be updated in light of the development of newer types of robots in recent years in the Japanese context.

The industrial robot is defined in ISO (the International Organization for Standardization) as '[a]n automatically controlled, reprogrammable, multipurpose manipulator programmable ... for use in industrial automation applications' (The International Federation of Robotics (IFR) 2015, my ellipsis). This definition refers to robots typically used in manufacturing industries. This definition has been recently expanded to add a new standard for the category of 'service' robot (The International Federation of Robotics (IFR) 2015). Service robots are regarded as machine devices that assist humans and that are used in non-industrial environments, such as personal household environments, or professional, institutional environments including hospitals or care facilities.

A parallel expansion of the definition of what a robot may be has taken place in Japan. For a long time, in the Japanese Industrial Standards (JIS), the robot was defined in terms of industrial robots. In recent years, non-industrial robots have been defined separately as '[i]ntelligent robots', '[m]obile robots', and '[s]ervice robots' (JIS Kikaku Yōgo [JIS Industrial Standard Terms website] 2016, original English). On the other hand, the Japanese government's Ministry of Economy, Trade and Industry (METI) defines the next-generation robot in terms of two spheres of activity: the new type of industrial robot that can work with humans and the service robot that coexists with humans, providing assistance and help in the areas of cleaning, security, welfare, care, or amusement (2005, 5).

There is a subtle difference between the OED and Japanese dictionaries such as Kōjien. While the meaning of the robot as automated machine is similar, the Japanese dictionaries describe a robot as automaton or artificial person and as a person who is controlled like a puppet (cited in Sena 2004, 9). The OED stresses that the robot is a machine. Taken together, the two definitions in the Japanese dictionary indicate the robot as both an artificial person and a person who is controllable. This ambiguous notion may be due to the fact that the term robot was first translated into Japanese in 1923 as '*jinzō ningen*', which means artificial human, when Čapek's *R.U.R.* was staged at *Tsukiji shōgekijō* (Tsukiji Little Theatre) in Tokyo in 1924 (Sena 2004, 35). It is indeed the case that Čapek's mass-produced labour robots are closer to 'clones' in our understanding, and, hence, the term 'robot' can be understood both as 'artificial human' and in terms of manipulable labour. Importantly, according to writer and journalist Timothy Hornyak, 'many Japanese found the concept of an artificially created human intriguing instead of terrifying, and jinzo ningen became a catchword' in the 1920s (2006, 34).

According to anthropologist Victor Turner, symbols are multi-referential: they can unify and condense diverse meanings and can also contain polarised meanings (1967, 29–30). Further, these meanings are culturally specific: the robot in its history has signified something frightening in the West while it does not necessarily convey this meaning in the Japanese context.[8] The symbolic and communicative power of a clone-like robot, already strangely attractive to a Japanese audience, is different from that of an automated machine. This type of a robot is a figure of ambivalence, human-like but not human. As we shall see, the longstanding Japanese concern with an animating of nature or the world of objects significantly colours the reception of robots in Japan.

The History of the Japanese Performing Robot

French inventor Jacques de Vaucanson and Swiss engineer Pierre Jaquet-Droz built automata in the late eighteenth century using period clockwork technologies of the early modern period that predated the development of Western automata. Scholars articulate the affective power of these automatons on their viewers. For example, historian Minsoo Kang provides detailed accounts of how the automaton has worked on 'the Western imagination, the object's capacity to arouse a wide spectrum of emotional reactions' from medieval times to the modern era (2011, 305).[9] Relevant to the current study, theatre historian Kara Reilly discusses the impact of automata on the audiences for which they performed (2011, 7). For Reilly, the automaton is conceived as a unique object for entertainment purposes, whereas the robot is regarded as a mass-produced machine substitute for the human labourer (2011, 176). These automata of the eighteenth century were performers rather than labourers.

Karakuri ningyō were, in effect, early entertainment automata that similarly capitalised on early modern Western clockwork technology. The term refers to a Japanese wooden automaton with clockwork-like parts inside and a spiral spring.[10] This Japanese style of automata became popular in the seventeenth century, a time of international isolation, when the Tokugawa shogunate prohibited foreign trade except for limited exchange with the Dutch and Chinese. The Japanese engineers of that time, drawing upon knowledge of Western clock-making, turned to the production of automata. *Karakuri ningyō* were not mass-produced dolls. They were uniquely made and associated with particular master *karakuri* craftsmen. Scholars have suggested that the tradition of *karakuri ningyō* entertainment has continued in the development of the next-generation Japanese robot (Umetani 2005, iii; Suzuki 2007b; Hotta 2008, 187; Asada 2010, 20). For Japanese anthropologist Masao Yamaguchi, the fact that *karakuri ningyō*, as 'charming copies of the human figure' that appeared 'in the world of entertainment, in a ludic ambience', contributed to 'tam[ing] [the fear and doubt regarding] the machine' for the Japanese (2002, 78–9, original English).[11] These automata were neither threatening nor did they evoke suspicion but were rather regarded as 'objects of curiosity in the milieu of spectacle' (Yamaguchi 2002, 79, original English). Roboticist Shigeki Sugano argues that the *karakuri ningyō* embraces artistic 'abstraction' as in the traditional Japanese performance forms Noh and Kabuki, rather than reflecting a 'realism' that has been more evident in the history of Western automata (2011, 36). Some *karakuri ningyō* embody a dramaturgy of the fallible as an entertainment strategy: for example, the automaton in the form of an archer that fails to hit a target with his arrow establishes an empathetic feeling toward the *karakuri* 'performer' (Suzuki 2007b, 44–5).

For Yamaguchi, the Japanese affinity for the robot has arisen as a function of the 'taming' of Western models of automaton technologies toward a model that was confined to entertainment (2002, 78). The tradition of the ludic automaton in Japan continued to the early twentieth century, while accommodating new technologies. In 1927, American electronics manufacturing company Westinghouse presented the Televox, a telephone-answering device that was also capable of turning the on-and-off switches of other devices. It was presented in 1928 in an anthropomorphised manner: the machine was fitted into the 'chest' of a roughly human-shaped figure. In the UK, Eric, a human-shaped robot with an aluminium body and bearing a plate with 'R.U.R.' on its chest, was presented at the Exhibition of the Society of Model Engineers in 1928. Eric stood up from a chair and delivered an opening speech via the transmitted voice of a human to the crowd, while twisting its head and raising its hands. The news about Televox and Eric was enthusiastically received at a time of growing popular interest in the robot following the introduction of *R.U.R.*, the play, in the mid-1920s in Japan (Nakayama 2006, 21; Kubo 2015, 57). The image of *jinzō ningen* became widespread through 'popular science magazines like Kagaku Gashō (Illustrated [sic] magazine of science)' and satirical writings in the 'robot boom' of that time (Nakamura 2007, 6, original English).

Biologist Makoto Nishimura created a Japanese version of a modern automaton, Gakutensoku (meaning 'learning of natural law'), and presented it at the exposition to celebrate the enthronement of Emperor Hirohito in 1928. It consisted of a human-height-sized upper torso and head on top of a decorative altar of approximately two metres in height. Through a pneumatic system with rubber tubes, Gakutensoku was able to move its face, eyes, eyelids, and arms, presenting smooth facial and gestural expressions. It enchanted visitors in a theatrical atmosphere, a Greek temple-like venue, with solemn music and an artificial singing bird. When a ceremonial mace in its hand was lit, this automaton, dressed in a toga, lifts its head, and its face was wreathed in subtle smiles (Aramata 1996, 28). According to writer and translator Hiroshi Aramata, Nishimura regards Gakutensoku as a creature with breath and warm blood, and some audience members took their hats off and offered their prayers (1996, 30). It is worthwhile noting that Gakutensoku is labelled as Japan's first robot (Nakamura 2007, 7; Aramata 1996, 26; Kubo 2015, 61).

The meaning and popularity of *karakuri ningyō* and Gakutensoku need to be examined in the Japanese context. *Karakuri ningyō* do not reference the creation of a human figure in 'the way in Europe automata design appear historically alongside alchemical efforts to create a homunculus and the Jewish mystical creation of the Golems' (Geraci 2010, 151). Nishimura regards Gakutensoku as nature's grandchild because humans are 'the children of nature' (MacDorman et al. 2009, 488). In the West, the idea of the automaton was to recreate the human being from dumb matter, while the Japanese vision is that humans are already a part of a lively nature, and so the robot is part of nature. This different understanding between the West and Japan continues to the current day. It is therefore understandable that Western writers (e.g., Schodt 1988; Hornyak 2006; Kaplan 2011) who have lived in Japan and are familiar with Japanese culture attribute the supposed Japanese affinity for the robot to the cultural tradition of animism, which see spirit incarnated in nature and in inanimate objects. Other important influences are Buddhist thought and robot manga/anime, which I will treat in more detail later. Japanese roboticists, writers, and critics argue the same points (Matsubara 1999; Sena 2001, 2004, 2008; Okuno 2002; Yonezawa 2002; Umetani 2005; Ishiguro 2007; Sugano 2011; Hotta 2008; Asada 2010; Kishi 2011; Masahiro Mori 2014).

Researchers in human-computer interaction Karl MacDorman, Sandosh Vasudevan, and Chin-Chang Ho argue that despite the sensational media coverage of the next-generation robots in Japan and their popularity, there is no conclusive evidence to suggest that 'Japan is a robot-loving society' (2009, 507). MacDorman, Vasudevan, and Ho suggest paying more attention to the pragmatic reasons for measuring the significance of robots in Japan, such as the country's post-war reliance on robotisation for manufacturing industries, the 'positive feedback' between robotic engineers' expertise and new markets, the secure position of in-house engineers in companies, and government initiatives and polices (2009, 504–5). In order to avoid the simple dichotomy of Japan as a robot-loving country and Western ambivalence toward robots, anthropologist Akinori Kubo, on the other hand, suggests a dynamic view of the robot in Japan. Kubo sees the robot as standing for the country's uncertain role in the modern world, negotiating Japan's imported modernity and the country's traditions (2015, 14). Kubo stresses that Japan's view of robots should be understood as an active response to the various stages of the country's modernisation

(2015, 14).[12] Kubo's view provides a larger framework for Japan's modernisation over the points raised by MacDorman, Vasudevan, and Ho, who look at the methodological difficulties in determining cross-cultural differences with regard to affinities for the robot (2009). However, Kubo's discussion can be read as a defensive description of the particularities of Japan's responses to modernisation through the figure of the robot, as he unwittingly generates another dichotomisation between Japan and the West.

In contrast, this book takes the view that cultural expressions regarding robots can be examined as complex responses to the institutional implementation of Japan's modernisation. I argue that next-generation robot projects constitute an important part of Japan's ongoing modernisation, led by the past and current governments. The Japanese attitude toward the possibility of domestic, daily contact with robots is still ambivalent, in part because 'they [Japanese people] are particularly concerned at [sic] the emotional aspects of interacting with robots' (Bartneck et al. 2006, 225).

While, at the moment, the Japanese may prefer their robots firmly located in the spaces of anime, manga, fantasy, and robot expos, situations that I will discuss in this book, it is important to see these not as comfortably separate entertainments but as products and producers of the social. An anxiety concerning possible emotional impacts points toward what I will suggest is an anticipation of relationships with robots, even if the circumstances for such relationships do not exist yet. I will further argue that the nature of this anticipated engagement is specifically Japanese as there is no strong opposition to the governmental and industry-led rhetoric of this engagement. Following Kubo, I suggest the robot represents 'positive feedback' between culture and technology within the history of Japanese modernisation. At the same time as the figure of the robot is used to navigate Japan's present and future, Japanese modernity should be understood as a 'cataclysmic modernity', meaning that it is 'a project compelled to draw on the traumatic memories of the past' (Eckersall 2013, 122). This darker, conflicted side of Japanese modernity runs underneath the surface of the popular events I will discuss.

What needs be highlighted is that there is an intrinsic connection between modernisation as it occurs in Japan and the particularities of Japanese cultural expression concerning robots. Andrew Feenberg, philosopher of technology, discusses the complex entanglements between technology, society, and culture in a globalised world (2010). Feenberg discusses the globalisation of technology in terms of 'branching' and 'layering'. The former denotes multi-directional spreading like a tree branching in which 'the trace of [Western] values appears clearly in design features of technical artifacts' in the new culture (Feenberg 2010, 123). In the 'layering' of technology, specific cultural values and codes, Japanese in this case, are also maintained with regard to technical objects. For Feenberg, Japanese modernisation presents a case for his argument that technology is 'no pure realization of rationality' and technological realisation is 'just as particular as any other expression of culture' (Feenberg 2010, 123). Japanese robots are such particular technologies.[13]

Japanese Modernity and Technology

Japan began to modernise during the Meiji Restoration in the late nine-teenth century, following United States Navy Commodore Matthew Perry's gunboat diplomacy in 1853 and 1854.[14] Japan's technological capacities were seen as inadequate and backward. In order to reconstruct Japan as a modern nation-state, the new government adopted the techno-logically deterministic slogan of *'shokusan kōgyō'* (promoting enterprise and expanding products) and *'fukoku kyōhei'* (enrich the country, strengthen the military), directing a rapid industrial development through Western mod-ernisation. In general, even though these policies were imposed from above, the public did not reject modernisation in its implementation stages (Murata 2003, 162). The effects of the technological and institutional modernisation by the Meiji government spread into social and cultural domains in Japan's growing, capitalist industrial society of the 1920s and the 1930s. Cultural modernisation was represented by the figures of the Westernised *mobo* (modern boy) and *moga* (modern girl), who gathered around European-style cafés or department stores and who were familiar with jazz music and ballroom dancing.[15]

This does not mean that there was no opposition to these changes. Japan's modernisation was promoted as Westernisation and industrialisation. According to historian Tetsuo Najita, powerful modern technology as 'a system of knowledge and production belonged to the Western Other' and 'its concomitant theories of positivism and progress' warped 'the ethical and aesthetic sensitivities of the Japanese' (Najita 1989, 11–2). Japanese culture was seen as polarised between its high culture, characterised by spiritual, fixed, still, eternal, and masculine values, and a low culture resulting from Western influences of the popular, vulgar, and transient, of mass consumption of the commodity, and it was associated with the feminine (Harootunian 2000a, 119 and 140).[16] Hence, modernisation was 'perceived as a threat to the "native tradition"', which prompted a traditionalist and nationalist backlash by political and cultural elites, paving the way for Japan's ultra-right-wing militarism of the 1930s and the 1940s (Starrs 2011, 110).[17]

The most notorious example of a nationalist reaction in the early years of the war was a two-day symposium in Japan entitled 'Overcoming Modernity' in 1942, seven months after the attack on the Pearl Harbour. The symposium was conceived as a vehicle for a reassessment of Japan's rapid modernisation and as an opportunity for Japanese intellectuals to 'consolidate their efforts for the good of the nation', by critically examining Western influences against the backdrop of the wars in Asia and the Pacific (Calichman 2008, xi).[18] The sense of the country's national crisis felt by participants was understood in terms of the jeopardising of Japanese culture and its venerated traditions (Calichman 2008, 18). Western influences since the Meiji Restoration were felt in both tangible and intangible ways. The cultural elites at the symposium were, however, unable to resolve what seemed oppositions between Japan and the West—the authentic and the inauthentic, or the inside and the outside—in terms of the question of cultural identity. Their efforts to solve the predicament of modern Japan in these terms could only end in a circular logic. For historian Harry Harootunian, the symposium merely returned to 'the place where Japan itself had been overcome by modernity' (2000b, 94). This dichotomy between Japan and the West is the unresolved conundrum at the core of Japanese modernism. The division is equally evident in Japan's understanding of its technology. The phrase, '*wakon yōsai*' (Japanese spirit and Western technology) represents the desire to solve the problem.[19]

There is a subtle difference in understanding the meaning of the concept of '*wakon yōsai*' before and after World War II. According to Najita, to counter the foreignness of modern industrialisation, a notion of timeless Japanese traditional culture and aesthetics was seen to provide implicitly anti-modern 'alternative [metaphysical] spaces to technology' in the pre-war period (1989, 15). While Japanese native culture was located in opposition to the technological other of the West in the pre-war discourse, in the post-war reconstruction phase, 'the bifurcation between culture and technology was mapped over in terms of progressive development' (Najita 1989, 13). Importantly, in the latter, technology is 'idealized in terms of "culture"': Japan's technological success, achievement, and excellence are thought to be 'an extension of cultural exceptionalism' (Najita 1989, 13). Japanese technological achievements in science and engineering were now promoted as central to Japanese cultural essence, as such. This is the 'techno-nationalistic discourse' of post-war Japan that sociologist Shunya Yoshimi discusses (1999, 151). In turn, Japanese technocentrism has more recently been seen through a lens of 'techno-Orientalism' that regards Japan as the technologised other (Morley and Robins 1995, 147–73), a matter I will take up in Chap. 5.

The figure of the robot is conceived at the centre of this historical tension between culture and technology. In debates on the next-generation robot, the supposed Japanese affinity for the robot is often discussed by proponents of robot technologies in a manner close to that of '*nihonjinron*', essentialist discussions about Japanese uniqueness or innate Japaneseness, in contrast to the West.[20] For example, next-generation robot projects are often discussed in terms of the history of Japanese craftsmanship, including the *karakuri ningyō*. Social psychologist Yotaro Takano points out the 'illusory correlation' between biases, validation, and familiar information in the circular logic of the *nihonjinron* argument (Takano 2008, 231). Cultural anthropologist Takeo Funabiki, on the other hand, states that *nihonjinron* surfaces when Japan faces a crisis, and, in the current context, the government's and the manufacturing industries' slogan for a return to Japan's traditions of craftsmanship should be understood in terms of the country's prolonged crisis due to the recession since the early 1990s (2010, 198–9). What I aim to highlight is how the rhetoric of the Japanese next-generation robots resonates with the technocentrism of Japanese modernity, its mobilising of Japanese cultural history and references for nationalistic and corporate aims.[21] The staging of Japanese technologies against the West for the Japanese audience, with ideological goals—just like Gakutensoku—has continued to the present, concerning anthropomorphic and zoomorphic robots. Enacting Japanese essence in a way that connects Japan's cultural past to its technologised future, these robots become performative objects that excite both the producers and their enchanted audiences.

Theatre and Performance Concepts Through Japanese Culture

Performativity and Geinoh

The term performativity is one of the most prominent theoretical concepts in the disciplines of theatre and performance studies. While it can be used generally 'to denote the performance aspect of any object or practice under consideration', basically, when referring to an action, it is a contested and elusive term (Loxley 2007, 140). Philosopher Judith Butler had famously developed, in her earlier work, her discussion of gender performativity by applying Victor Turner's view of the repetitive performance of ritual social drama as both re-enactment and re-experiencing (Butler 1988). When I discuss 'performativity' in relation to Japanese mimetic robots, I am referring to a generative and signifying capacity, not a subversive potential, as in Butler's sense of the term. What I will argue throughout this book is that the performativity of Japanese robot performance events is evident through an evocative iteration of representation and expectation within a culturally specific context, one that is facilitated by the transformative effects of particular *mises en scène*.

The main performers of the robot-themed events in this book are the mimetic robots that figure in the chiasm between the real and representation. Robots' resemblances to manga/anime robot characters, as well as to real humans and animals, cannot be discussed in terms of their biomechanics and technical facility alone, the logic of which stresses that robotic mechanisms can expertly replicate the natural mechanisms of living creatures, such as animals, birds, fish, or insects, or that it is possible to make a robot that literally actualises a manga character.[22] I am concerned with the representational effects of robots and their contextualisation, a matter that is not discussed in science or engineering contexts. In this sense, the robot as a performing object has the capacity to implicate participants in its generation of meaning through its performance in a culturally specific, staged event or situation.

Performance studies, as articulated by influential scholar Richard Schechner, seeks to address the complexity of contemporary living in relation to aesthetic practice, covering social, ritual, and quotidian activity within its scope (2013). Its broader view of performance encompasses an examination of the 'everyday'. In more specific terms, I follow German performance theorist Erika Fischer-Lichte, who deploys 'an expanded definition of performance beyond Western institutionalized theatre, in order to include a range of popular entertainments, rituals, and political and communal interactions between people across a variety of cultures' (2014, 16). I apply this inclusive understanding of 'performance' to the robot in the Japanese context.

In Japanese language, the word '*engeki*' is used to translate 'theatre' in English. However, it strictly designates modern theatre forms. For traditional performances such as Kabuki, the term '*shibai*' is used. According to ethnologist and folklorist Shinobu Orikuchi, there is also a more general term, '*geinoh*'—originating from '*matsuri*' (festival or carnival) activities at the site of '*kyōen*' (banquet or banquets) since Japanese antiquity—which can be used to refer to much wider cultural activities or public entertainments (1991, 19). For Japanese studies scholar Jacob Raz, *kyōen* means a participatory event for which everyone is expected to 'perform something' whether it is a dance, a song, chanting, or poetry reading (1983, 28). According to Orikuchi, there was no concept of the 'audience' as such in pre-medieval Japan, as everyone participated, and as curious gods were presumed to regularly come to watch *matsuri* as uninvited guests (1991, 36). Historian Ian Buruma explains that '[p]ain and ecstasy, sex and death, worship and fear, purity and pollution are all vital elements in the Japanese festival' (1984, 11). *Matsuri* in essence allow temporary transgressions of rules, conventions, and boundaries, just like the carnival in the West.

I discuss robot demonstrations and competitions in terms of *geinoh* (and *matsuri*), highlighting their underlying theatricality. The term 'theatricality' is another contested term within theatre and performance studies, which requires care when using it in a Japanese context. The term is associated with traditional theatre terms such as mimesis and the *theatrum mundi* with its paired oppositional concepts such as real and false, genuine and counterfeit, original and copy, or honest and dishonest (Davis and Postlewait 2003, 2–10). The concept of theatricality has been criticised as limiting for discussion of performance works outside traditional plays, such as modernist avant-garde performance, performance art, or post-dramatic theatre as well as in relation to cultural performance such as ritual, carnival, or sports events. Theatre studies scholar Janelle Reinelt points out, however, that the narrower use of the term 'theatricality' is not universal, and there are cross-cultural misunderstandings of the term (2002, 209). For example, the above sense of 'theatricality' does not apply in German and French contexts, as the European notion of the term is not confined to traditional theatre (Reinelt 2002, 208).

In the Japanese context, theatre studies scholar Mitsuya Mori discusses the particular deployment of theatricality in *geinoh*, which includes the Japanese traditional performing arts forms, such as Kabuki and Noh, as well as cultural activities such as the 'tea ceremony and *Sumo* wrestling' (2002, 90). Mori stresses that the common Western notion of theatricality arising from the triangulated relationship of actor, character, and audience, as in realistic theatre—which is distinguished from the other Western triad of player, role, and spectator—is not applicable to Japanese performance traditions that actively engage with and present the 'intersection of reality and fictionality' in performance (Mitsuya Mori 2002, 90), thus blurring boundaries between audience and performer. Mori's discussion, however, engages with the 'closed' performances of theatre practice, as it addresses an expert audience with '"correct" reception', and is not concerned with what theatre semiotician Marco De Marinis terms 'open' performance that addresses 'a receiver who is neither too precise, nor too clearly defined in terms of their encyclopedic, intertexual, or ideological competence' (1987, 103). This book looks into robot events as both 'open' and 'closed', meaning both public performances and those for initiates and aficionados, like robot competitions. I apply the Japanese notion of *geinoh* and this notion of an open *matsuri* participation to robot entertainment performances.

The Performing Object and Staging

The concept of 'performing object' is commonly used by puppetry scholars and practitioners to designate objects on stage such as puppets or masks for performance.[23] In addition, puppeteer and scholar John Bell includes objects such as sculpture, painting, and ritual objects in performance, and he refers to the automatons of the eighteenth and nineteenth century (2008, 2). Bell also suggests an expanded understanding of performing objects that may include 'computers, video, screens, film screens, telephones, [and] radios', and he discusses how these machines 'perform with us and for us', transmitting 'stories and ideas' (2008, 3). This enlarged notion of the performing object highlights the significance of multi-layered relationships with everyday objects, and that these objects seem to have an 'ability' to engage with people (Bell 2014a, 9).[24] The design philosophy for interactive mimetic robots intended for the domestic sphere is to establish similar relationships. I treat all of the robots discussed in this book in terms of such engagements, as performing objects.

Bell remarks that the performing object can also be associated with primitivism and children's entertainment. The animistic nature of the puppet, masks, and other performing objects, which acknowledge a kind of spiritual agency or infusion of vitality, is regarded as antithetical to modern, rational thinking (Bell, 2008, 6). In this regard, Bell states that the Freudian notion of the uncanny—in which something familiar or homely is repressed and resurfaces as the 'unhomely'—is used in discussion 'to tame the effects of object theatre by assigning them to the irrational and pathological, rather than to consider the disconcerting possibility of the agency of things' (2014b, 49). Bell is in agreement with Kang, who discusses the European imagination concerning the automaton when critiquing the notion of the uncanny. Kang argues that the idea of the Freudian uncanny, which has negative connotations, is not sufficient for a discussion of the automaton that can arouse in the viewer complex emotional responses—such as enjoyment and captivation, as well as a sense of eeriness, or of being startled (2011, 22–8). As discussed above, animistic thinking remains alive in Japan. How might a performing object in the Japanese context, particularly the robot, be regarded, and is it uncanny? In order to consider this question, I refer to Kang's discussion.

In articulating the viewer's feelings when encountering the automaton, Kang critically discusses roboticist Masahiro Mori's concept of 'the uncanny valley' (2011, 47–51). Mori is a pioneering figure in the history of Japanese robotics. In 1970, Mori hypothesised that as machines appear more human-like, '*sinwakan*' (meaning familiarity, affinity, or rapport), people's sense of liking and connection with them increases until at a certain point there is a sudden drop, a 'valley' when robots become creepy (Masahiro Mori 2012, 99). But, as robots increasingly resemble humans, the positive feeling toward them then increases, and at the highest level of resemblance, the perceiver's connection with the robot is like an affinity for a healthy human. As one of his points of comparison, Mori situates the Japanese Bunraku puppet on the positive side of this equation. Kang criticises Mori's reference to the Bunraku puppet as problematic and the question of its uncanniness as irrelevant, as Bunraku operates in a theatre context that incorporates distance from the audience (2011, 47). However, Mori's comments on the Bunraku puppet, in fact, highlight the transformative effects of Bunraku theatre. Mori states that when members of an audience become absorbed, they 'might feel a high level of affinity for the puppet' even though '[i]ts realism in terms of size, skin texture, and so on, does not even reach that of a realistic prosthetic hand' (Masahiro Mori 2012, 98). What is missing in

Kang's account is a recognition of the theatricality of the staging that would have surrounded the automaton, and certainly frames the Bunraku puppet. Representation is consciously deployed in carefully crafted spaces to be perceived and read by the audience in particular ways. I argue that mimetic robots similarly need to be considered in terms of a dynamic and culturally specific *mise en scène*.

DOUBLE VISION

Using the term 'double vision', Steve Tillis, a theorist of puppetry, discusses the double function of the puppet, in which 'the audience see the puppet, though perception and through imagination, as an object and as a life; that is, it sees the puppet in two ways at once' (1992, 64). Similarly, Fischer-Lichte states that the spectator might start seeing objects on stage as a 'sign-bearer that is linked to associations – fantasies, memories, feelings, thoughts – to what it signifies, i.e., possible meanings' (Fischer-Lichte 2014, 39). The complex and paradoxical mode of theatrical perception is well summarised by French phenomenologist Mikel Dufrenne, when he said: '[in theatre] I do not posit the real as real, because there is also the unreal which this real designates; I do not posit the unreal as unreal because there is the real which promotes and support this unreal' (quoted in Power 2008, 189). The audience of puppet theatre actively engages in 'make-believe', a paradoxical modality in which an object can appear to have a life through representation (Walton 1990, 11). I would argue that the 'audience' for robot events and situations participates in that engagement in a similar way. I remarked earlier that the robot is both a real object that is the focus of advanced scientific development, and a powerful trope within the Japanese social imaginary. I am interested to explore this notion of seeing the real in the unreal (which is also to see the unreal, meaning representation, in the real).[25]

The *mise en scène*, as I've indicated for the effects Bunraku produces, plays an essential role in this 'double vision' where mimetic robots are concerned. Robots' existence between reality and representation in Japan is predicated upon its highly coded contexts as they are interpreted by the audience in question.

Participatory Spectatorship

Scholars of the puppet and the automaton emphasise the importance of the audience in discussing the performing object.[26] For example, Dassia Posner states that '[a]nalyses of performing-object theatre … must consider …

multiple modes of experience that cross over, diverge, and harmonize in the mind of the viewer' (2014, 226, my ellipses). So, the question arises of what elements are present, and in what ways, and how they create this multiplicity of experiences that is synthesised by the audience. These include emotive effects and affects. One of the key objectives of Kang's study is to articulate the automaton's 'ability to arouse powerful, conflicting emotions in people' (2011, 5).

What kind of audience engagement occurs at a robot competition or with a domestic robot in the home? This book will discuss the conscious engagements by the audience when participating in Japanese robot events and situations. They are willing participants of the game of 'what if'—being aware of the doubleness of 'the "spectating I" and the "I of the spectator"' in theatre studies scholar Peter M. Boenisch's sense—and enjoy the transformations of the robot in the performance situation, knowing that it is not real (but also real) (2014, 237). They are not, however, 'the emancipated spectator', as in French philosopher Jacques Rancière's sense, who is an autonomous subject playing 'the role of active interpreters, who develop their own translation in order to appropriate the "story" and make it their own story' (2009, 22). The type of spectatorship this book explores is not one of a critically liberated, independent viewer, as Rancière advocates, although spectators may interpret robot performances as they will. According to anthropologist Rupert Cox, '[p]layful activities are not "play" because they suspend the rule of reality, but because those who participate accept the context which constrains their action and the ludic structure which frees it' (2002, 182). The spectatorial engagements that this book discusses follow this ludic logic: they are simultaneously rule-bound, along several axes, those of the situation itself and those governing the social imaginary that have already shaped the preconditions for the interaction (or determined its outcomes, in the case of government or corporate events), and may be facilitative of playful creativity within given frameworks established for the robot-human exchange.

For a discussion of the complex nature of Japanese spectatorship at robot performance events, it might be necessary to consider the notion of the 'passive' audience as one modality in which audiences may engage with robots. Here, I refer to applied theatre and community theatre practitioner and theorist Gareth White's discussion on audience participation (2013). White discusses a different reading of the conventional spectator, where 'an encounter with a work of art is thought of as giving oneself over, to be played by the work, rather than to have control over it' (2013, 186–7). For

White, though this encounter can be seen as passive, 'one is played by it, and becomes something different because of it, a change that is available to reflection after the event' (2013, 187). What is most useful in White's thesis is his awareness of both aspects of 'heteronomy and autonomy' in regard to the spectator, stressing 'the continuity of the participant's social being, and how it is connected to and marked off from an altered version of itself in performance' (2013, 206).[27] White highlights the transformative aspect of theatrical experience, which allows 'us to perceive ourselves anew' (2013, 206). Similarly, Fischer-Lichte argues for the agency of the spectator as a key component of performance, with the German term '*Aufführung* (live performance)' that emphasises a transformative power of the encounter of spectators and performers (Arjomand and Mosse 2014, x).

An emphasis of the connection between the components of the production and the audience suggests a more nuanced understanding of *mise en scène*. For example, for Fischer-Lichte, *mise en scène* brings forth a viewing structure for a performance 'in such way that the appearing elements attract the audience's attention and simultaneously highlight the very act of perceiving itself' (2008, 188–9). Through a particular *mise en scène*, the spectators are made conscious of being affected by 'the movements, light, colors, sounds, odors, and so forth' (Fischer-Lichte 2008, 189).

The heightened connection that arises in the dynamic relation between the totality of the piece and its audience through a particular *mise en scène* can be discussed in terms of the Japanese notion of '*ba*', which contains simultaneous meanings of place, spot, space, room, occasion, situation, scene, or field. The complex meaning of *ba* can be understood through philosopher Kitaro Nishida's discussion of '*basho*' (place), which is often used interchangeably with *ba*. Nishida's philosophy highlights the importance of *basho* in a Japanese context as the essence of an understanding beyond the subject-object division (2007, 429). *Basho* is understood as a structure in which 'each actor "negates itself" to become the "world" for the others, that is, the place of the interaction' (Feenberg 2010, 116). Each encountering party becomes an encompassing and environmental medium for the other to mediate and transform. Japanese roboticists refer to the term '*ba*', and in their usage the term strongly suggests this underlying meaning of mutual mediation and transformation though this important notion is not often remarked upon in Western discussions of Japanese robotics (Okada 2012, 24–5; Takeuchi 2007, 266).

Participants in the traditional *matsuri*, including audience members, are required to be as aware of their actions as actors in a play. As I have mentioned, in pre-modern Japan, *matsuri* were staged for the entertainment of *marebito*, gods from foreign lands or messengers from the afterlife, with food, drink, dancing, and singing through *kyōen* (Orikuchi 1991, 22). The *matsuri* actors perform between the sacred and profane, or the everyday and the non-everyday, for the duration of *matsuri*. Playing an audience member means taking part in the inclusive *mise en scène* of the *matsuri*. The audience becomes, per the concept of *ba*, an experiential medium for the *matsuri* performers. Importantly, the mutuality of this 'acting' can be found in contemporary, mundane 'rituals' in Japanese society, such as in the drinking ritual, deriving from the traditions of the *kyōen* banquet, of pouring each other's sake, and thus being simultaneously a host and a guest (Toida 1994, 86).

I am saying that the Japanese audience comes to robot performance with an acculturated knowledge of the conscious participation and immersive nature of the *matsuri*, which, as I have indicated, can also inflect contemporary, quotidian social activity. Magico-religious operations and effects, though not so much meanings, have filtered through to the everyday social such that ordinary Japanese people can easily appreciate non-religious robot performance events as *geinoh* through an established tradition of festive make-believe. The pleasure of viewing/participating is the very ambiguity of the audience's viewing position, allowing a space for contradiction, paradox, or fakery that is also a suspension of resolution of meaning between useful and useless, authentic and inauthentic, or real and copy.

In This Book

This book is a survey of a range of expressions of robot culture in Japan at the beginning of the twenty-first century. The case studies of robot performances I discuss, at a time of hype regarding next-generation Japanese robots, include demonstration shows, competitions, fan gatherings, tourist entertainment, and monuments. These diverse expressions are episodically connected, as for Deleuze and Guattari's model of the rhizome, to form an associational map of the contemporary Japanese imaginary, shaped by traditional culture, institutional aspirations, technologists' dreams that are themselves drawn from the lexicon of post-war futurism, and the dark perspectives of a modernity marked by trauma and frustration. The case studies from Chap. 3 onward are arranged in a way that moves from the public context to the domestic (or at least as the domestic is referenced in the nursing home).[28]

In Chap. 2, I examine writings by Japanese roboticists and cultural theorists who discuss anthropomorphic bipedal robots in Japan. In their writings, they highlight how popular-culture representations of the human-oid robot—that is, in manga/anime—have contributed to their stated rationales for robot development in its early phases. This chapter also discusses how the image of the robot in manga/anime lends a particular presence to anthropomorphic robot prototypes that renders them familiar and unthreatening. The aim of this chapter is to indicate the affective power of the appearance of next-generation humanoids in relation to these popular-culture sources and in the context of theatre and performance studies approaches.

Chap. 3 brings a critical perspective to government and corporate robot spectacles for the public. This chapter examines the nationalistic and pro-motional narratives in demonstration performances of robots presented as entertainment at large international events held in Japan since the 1970s. I highlight the rhetoric of robots as emblems of futurity that pervades these spectacularised robot events, which feature fantasies of future human-robot interaction, modelled for the wider Japanese public. I examine in what ways the dramatic and dazzling *mises en scène* of such events have been crafted to make the next-generation robots symbolically meaningful to the Japanese audience.

Extending an investigation of the affective *mise en scène* constructed for mimetic robots, Chap. 4 examines what I discuss as liminal aspects of robot performances in art productions. This chapter examines collaborative the-atre experiments by Hiroshi Ishiguro and Oriza Hirata and Ishiguro's android performance. These works are contrasted with artist Kenji Yanobe's installation work of a fire-breathing, gigantic robot doll presented in a visual arts context—a work that reveals the robot as uncanny or strange. In contrast to Chap. 3's triumphalist expressions of Japanese culture through the figure of the robot, these artists' works offer more ambiguous and complex representations that dramatise differences between humans and robots.

Continuing to explore a range of contexts for affective engagement with performing robots, Chaps. 5 and 6 look into subcultural domains, a per-spective that rarely attracts critical attention by scholars concerned with social robotics. Chap. 5 discusses varied forms of fighting humanoid per-formance in popular entertainment. Though designed for different audi-ences (Japanese robot fans, and foreign tourists in the case of the Robot Restaurant), this chapter exposes how these popularised productions embody fantastical narratives to do with the image of a futuristic but wacky 'techno-Japan'. The robots in this particular prismatic reflection of the popular understanding of the robot are not nationalistic boosters, as they are in Chap. 3, or intended for the purpose of artistic exploration, as discussed in Chap. 4, but instead are humorous or outrageous performers of phantasmagoria for their fans. Importantly, spectators, creators, and performers knowingly and enthusiastically participate in this spectacle— the sort of demonstration that is an essential part of fan and tourist cultures.

Chap. 6 discusses a popular Pygmalion fantasy in the Japanese context, that is, the creation of an ideal woman, by examining the virtual singing performance of Hatsune Miku. Hatsune Miku is software that combines synthesised singing with an illustrated girl character. For its male fans, Hatsune Miku embodies the metonymical aural and visual signs of femininity. Fans within the *otaku* subculture manipulate and control the figure and its outputs through this software and related packages, which can generate images and music videos featuring the character. As scholars indicate, these productions can actually stimulate sexual arousal for hard-core fans. This chapter explores the darker side of the modernist fantasy of technologised progress by considering the figures of the robot and the woman in terms of themes of control and subjugation within the *otaku* subculture.

Robot competitions are an important aspect of the popular apprehension of Japanese robotics, yet there is little critical study of them. Extending discussion of hobbyist participation in robot culture in Japan, Chap. 7 looks into humanoid competitions. These competitions are organised events for purpose-built, small-scale hobby robots: they sprint, play soccer, or fight each other. Focusing on the fighting robot format, this chapter discusses these robots as important intermediaries, facilitating a reflexive anthropomorphism in which participants project themselves into their robots. Through an examination of a comedy contest for robots that is designed to entertain the audience with laughter, the chapter also examines the intersection of humour, robotics, and spectacle. As this comedy robot contest shows, humour can successfully facilitate the audience's imaginative capacity to accommodate something imperfect, strange, or unfamiliar, such as a humanoid robot.

Taking ideas of the theatrical encounter in the Japanese context into situations of close physical proximity, Chap. 8, critically examines the assumptions that have guided the development of robots designed to interact with the elderly. The use of the 'social' robot for aged care is highly calculated, designed so that users will adapt to and align themselves with the robots in the context of games and prompted conversations that facilitate the users' responses. It also raises ethical issues. Researchers are aware that the users may not be fully aware, depending on their health status, of the nature of the solicitous object. The final part of this chapter examines performative parallels between manga/anime that include narratives of robots used in aged care and related occurrences in real situations. These correlations suggest a deep social entanglement with robots in Japan, even in the conceptualising of aged care.

The book's epilogue examines the theme of the robot in relation to the 2020 Tokyo Olympics. The robot as a symbol of Japan's futurist ideology takes centre stage again, in the context of another national event (as *matsuri*) that seeks to represent Japan through the figure of the robot.

NOTES

1. Zoomorphic social robots include robot dog AIBO by Sony, baby seal robot Paro by AIST, and Omron's robot cat NeCoRo.
2. Exemplar humanoids include ASIMO by Honda and QRIO by Sony. Kokoro company's Actroid is a notable female android.
3. Social robotics has also developed in other countries. For example, notable roboticist Cynthia Breazeal's Kismet and Leonard at The Massachusetts Institute of Technology are desktop autonomous social robots that have gestural and facial expressions. Social robots are also not just research platforms developed by universities and corporations but can include varied commercial products. Toy robot products developed commercially in the USA include Furby by Tiger Electronics in the late 1990s and Hasbro's My Real Baby and Mattel's Miracle Moves Baby in the early 2000s. These robots were highly sophisticated interactive machines. French company Aldebaran Robotics produced the interactive and personalisable robot NAO in 2006. Japanese products include table-top communication robot PaPeRo, by NEC; static conversation robots ifbot and Hello Kitty Robo by Ifoo; and Robi, developed by famous robot designer Tomotaka Takahashi, which can walk, dance, sing, and converse with humans.
4. I use the combined term 'manga/anime' throughout this book unless a specific work is discussed.
5. The examples include Mark Pauline's Survival Research Labs, Chico MacMurtrie's Amorphic Robotic Works, Bill Vorn, and Louis-Philippe

Demers. Artist and theorist Eduardo Kac, on the other hand, discusses the role of robotics in performance art, including the works of Stelarc and Marcel.lí Antúnez Roca (2005).

6. I use Japanese names in this essay in the English way: given name first, followed by family name. Long vowel sounds are indicated by macrons, unless the word is in common usage in Romanised form (e.g., Tokyo not Tōkyō). However, I refer to the name without diacritical marks if a particular author's name is spelled in English without them (e.g., Shozo Omori not Shōzō Ōmori). Similarly, I do not use the spelling 'Ohmori' for Omori.

7. Sociologist Lewis Yablonsky uses the term 'robopath' to illustrate 'people whose pathology entails robot-like behavior and existence' and who are egocentric and 'socially dead', lacking compassion for others (1972, 7). Robopaths are the opposite of the robot, embodying machine-like functionality in 'contemporary social machine societies' (Yablonsky 1972, 31).

8. When I refer to 'the West', I mean European countries, Canada and the USA, and Australia and New Zealand. Because they are commonly used in Japan, terms such as 'the West' and 'the Western' are used in this book.

9. The idea of the artificial being and associated technologies has featured in the human imagination for millennia and can be traced back to Greek mythology and 'the legends of an immemorial past' (Cohen 1966, 15). Greek mythological narratives include those of the blacksmith god Hephaestus who created golden damsels; the proto-engineer Daedalus, who built statues that could move; and Pygmalion, a legendary sculptor who become obsessed with his own carved-ivory woman. In the fourth century BC, mathematician and philosopher Archytas is believed to have built a steam-powered bird. Hero of Alexandria, a Greek inventor and mathematician of the first century AD, is said to have built automata. In Europe, the idea of creating a living human statue was manifest in the form of the homunculus and the Golem. Paracelsus, Swiss physician and philosopher, regarded the homunculus as 'more important than the alchemical synthesis of gold' in Renaissance Europe (Geraci 2010, 154). The Jews in Europe sustained the myth of the Golem, a creature animated from mud and clay, from the seventeenth to the nineteenth centuries as a 'hope for magical aid against oppression'. It

was a creature that was powerful, yet inferior to humans (Geraci 2010, 156).

10. There are many kinds of *karakuri* automatons, including archers, letter-writers, and music-box *karakuri*. The most well-known style of *karakuri* is a tea-carrying doll that would walk (roll) toward a person while carrying a teacup. When the cup is lifted, it stops. When an empty cup is placed on the doll's tray, it turns and moves back to its original position.

11. When I quote a Japanese author's writing in English, I indicate it as 'original English'. Otherwise, all the quotes in English from Japanese sources are my translation.

12. According to cultural anthropologist Takeo Funabiki, Japan's sense of uncertainty and marginality about its identity comes from the fact that the country did not participate in the West's history of modernity (2010, 39–41).

13. Japanese technologies has been, from time to time, described, negatively, with reference to The Galápagos Islands, where their isolated environment nurtures its own unique evolutionary process of animal and vegetation species. 'Galapagosization' in the Japanese context, generally, means that Japanese electronic products are too specific and expensive to be global items, and remain popular only within Japan. There is also a positive take of the term, as it allows a unique development of the next-generation robotics (Numata 2001, 145). This book highlights a 'Galapagosization' that has created unique cultural performance practices through the figure of the robot.

14. The Meiji Restoration in 1868 re-introduced imperial rule to Japan under Emperor Meiji. It replaced the Tokugawa Shogunate, ending the Edo period, which commenced in 1603. Japan's modernisation, however, did not happen overnight after the Meiji revolution. The Tokugawa government had already begun to modernise in the nineteenth century (Murata 2003, 160; Starrs 2011, 122–3).

15. For a study on the mass culture of this period, see Silverberg (2006).

16. In the post-war period, writer Yukio Mishima famously assesses Japan in gendered terms, stating that the country is being feminised and emasculated since the Meiji modernisation (Starrs 2011, 238). For a discussion on a feminised Western modernity that is aligned

with modern mass culture and consumer society, see Huyssen (1986, 44–62).

17. For a discussion of the interrelations between fascism, militarism, and capitalism and Japanese intellectuals' responses to these in the inter-war period, see (Harootunian 2000b).

18. It was organised by Katsuichiro Kamei, with Tetsutaro Kawakami and Hideo Kobayashi, fellow intellectuals in the circle around the journal *Bungakukai* (Literary World). The participants selected were leading figures in literary criticism, film criticism, philosophy, theology, science, music, novel, and poetry, representing the scope of contemporary Japanese intellectual life. This event exemplifies the mobilisation of artists and intellectuals at a time of great national significance in Japanese history. A later example is the Japan Expo in 1970, which I discuss in Chap. 3. Similarly, the 2020 Tokyo Olympics is discussed in Chap. 9.

19. The phrase, '*wakon kansai*' (Japanese spirit and Chinese technology), was used for centuries. Such phrases reflect the underlying sense of anxiety and fragility regarding Japan's position in relation to China, or the West, that 'Japan's history is suffused with the sense of the dominant Other and its own marginality' (Miyoshi and Harootunian 1989, xi). Yet, comparative literature scholar Sukehiro Hirakawa stresses that Japan's eclecticism is such that Buddhism and Confucianism are naturalized, and '*wakon*' in Meiji Japan included these foreign philosophies (2006, 56).

20. Sociologists Yoshio Sugimoto and Ross Mouer point out the lack of empirical evidence in *nihonjinron* discourses (1995).

21. Selma Šabanović, scholar in Science and Technology Studies, points out that Japanese robotics researchers refer to 'untested and unquestioned cultural assumptions' to justify and normalise the new technologies they develop, and that there is a need for social science studies to reveal these suppositions (2014, 361). This book offers a critical reading of Japanese robotics researchers' assumptions from a theatre and performance studies perspective.

22. Highlighting biomechanical reasons, roboticist Koichi Suzumori discusses why the appearance of robotic machines resembles those of animals, birds, fish, or insects (2012).

23. The discussion of performing objects can also include ventriloquism (Tillis 1992; Shershow 1995; Connor 2000; Nelson 2001; Goldblatt 2006).

24. Recent studies on objects and material culture discuss human-object relationships from multidisciplinary perspectives. See, for example, Bennett (2010) and Turkle (2007).

25. Importantly, the paradoxical duality between the real and the unreal is characteristic of Japanese animism, as I will discuss in a later chapter. Referring to animistic beliefs in Japan, Japanese literature and theatre scholar Cody Poulton explains that 'if the uncanny is the disconcerting byproduct of the mimetic instinct in Western art', the Japanese accommodate 'the otherworldly' through Shinto animism (2014, 291). Poulton stresses that the Japanese can accept non-human otherness 'while still acknowledging and even celebrating its essential strangeness' (2014, 291).

26. In theatre studies, reception and audience form a complex area of ongoing investigation (Blau 1990; S. Bennett 1990; McConachie 2008; Kennedy 2009; Grehan 2009; Fensham 2009; Oddey and C. White 2009; Freshwater 2009; Radbourne et al. 2013; Heim 2016).

27. White's view of the duality of the spectator's autonomy and heteronomy echoes performance studies scholar Shannon Jackson's discussion of social art work, highlighting 'the contingency of any dividing line between autonomy and heteronomy, noticing the dependency of each on the definition of the other' (2011, 29).

28. I would like to thank Meredith Morse for this methodological approach, which she had deployed in a chapter of her book on US artist Simone Forti (2016). The chapter concerned brings the reader closer and closer to the workings and affects of the performing body.

References

Allison, A. (2006). *Millennial monsters Japanese toys and the global imagination.* Berkeley: University of California Press.

Aramata, H. (1996). *Daitōa kagaku kitan* [Mysterious science stories in Greater Asia]. Tokyo: Chikuma Shobō.

Arjomand, M., & Mosse, R. (2014). Editors' preface. In M. Arjomand & R. Mosse (Eds.), *The Routledge introduction to theatre and performance studies* (M. Arjomand, Trans.) (pp. viii–ix). London/New York: Routledge.

Asada, M. (2010). *Robotto to iu sisō* [A philosophy called the robot]. Tokyo: NHK Shuppan.

Auslander, P. (2006). Humanoid boogie: Reflections on robotic performance. In D. Krasner & D. Z. Saltz (Eds.), *Staging philosophy: Intersections of theater, performance, and philosophy* (pp. 87–103). Ann Arbor: University of Michigan Press.

Bar-Cohen, Y., & Hanson, D. (2009). *The coming robot revolution: Expectations and fears about emerging intelligent, humanlike machines.* New York: Springer.

Bartneck, C., Suzuki, T., Kanda, T., & Nomura, T. (2006). The influence of people's culture and prior experiences with Aibo on their attitude towards robots. *AI & Society, 21*(1–2), 217–230. doi:10.1007/s00146-006-0052-7.

Bekey, G., Ambrose, R., Kumar, V., Lavery, D., Sanderson, A., Wilcox, B., Yuh, J., & Zheng, Y. (2008). *Robotics: State of the art and future challenges.* London: Imperial College Press.

Bell, J. (2008). *American puppet modernism: Essays on the material world in performance.* New York: Palgrave Macmillan.

Bell, J. (2014a). Omnipresence and invisibility: Puppets and the textual record. In D. N. Posner, C. Orenstein, & J. Bell (Eds.), *The Routledge companion to puppetry and material performance* (pp. 7–10). London/New York: Routledge.

Bell, J. (2014b). Playing with the eternal uncanny: The persistent life of lifeless objects. In D. N. Posner, C. Orenstein, & J. Bell (Eds.), *The Routledge companion to puppetry and material performance* (pp. 43–52). London/New York: Routledge.

Benford, G., & Malartre, E. (2008). *Beyond human: Living with robots and cyborgs.* New York: Forge Books.

Bennett, S. (1990). *Theatre audiences: A theory of production and reception.* London/New York: Routledge.

Bennett, J. (2010). *Vibrant matter: A political ecology of things.* Durham: Duke University Press.

Berghaus, G. (2005). *Avant-garde performance: Live events and electronic technologies.* Houndmills/Basingstoke/Hampshire/New York: Palgrave Macmillan.

Birringer, J. H. (2008). *Performance, technology, and science.* New York: PAJ Publications.

Blau, H. (1990). *The audience.* Baltimore: Johns Hopkins University Press.

Boenisch, P. M. (2014). Acts of spectating: The dramaturgy of the audience's experience in contemporary theatre. In K. Trencsényi & B. Cochrane (Eds.), *New dramaturgy: International perspectives on theory and practice* (pp. 225–241). London/New York: Bloomsbury.

Broadhurst, S. (2009). *Digital practices: Aesthetic and neuroesthetic approaches to performance and technology.* Houndmills/New York: Palgrave Macmillan.

Brown, S. T. (2010). *Tokyo cyberpunk: Posthumanism in Japanese visual culture.* New York: Palgrave Macmillan.

Buruma, I. (1984). *Behind the mask: On sexual demons, sacred mothers, transvestites, gangsters, drifters and other Japanese cultural heroes.* New York: Pantheon Books.

Butler, J. (1988). Performative acts and gender constitution: An essay in phenomenology and feminist theory. *Theatre Journal, 40*(4), 519–531. doi:10.2307/3207893.

Calichman, R. (2008). *Overcoming modernity: Cultural identity in wartime Japan.* New York: Columbia University Press.

Causey, M. (2006). *Theatre and performance in digital culture from simulation to embeddedness.* London/New York: Routledge.

Cohen, J. (1966). *Human robots in myth and science.* London: George Allen & Unwin.

Condry, I. (2013). *The soul of Anime: Collaborative creativity and Japan's media success story.* Durham: Duke University Press.

Connor, S. (2000). *Dumbstruck: A cultural history of ventriloquism.* New York/Oxford: Oxford University Press.

Cox, R. (2002). Is there a Japanese way of playing? In J. Hendry & M. Raveri (Eds.), *Japan at play: The ludic and logic of power* (pp. 169–185). London/New York: Routledge.

Davis, T. C., & Postlewait, T. (2003). Theatricality: An introduction. In T. C. Davis & T. Postlewait (Eds.), *Theatricality* (pp. 1–39). Cambridge/New York: Cambridge University Press.

D'Cruz, G. (2014). 6 things I know about Geminoid F, or what I think about when I think about android theatre. *Australasian Drama Studies, 65,* 272–288.

De Marinis, M. (1987). Dramaturgy of the spectator (P. Dwyer, Trans.). *TDR/The Drama Review, 31*(2): 100–114. doi:10.2307/1145819.

Dery, M. (1996). *Escape velocity: Cyberculture at the end of the century.* New York: Grove Press.

Dixon, S. (2007). *Digital performance: A history of new media in theater, dance, performance art, and installation.* Cambridge, MA: MIT Press.

Eckersall, P. (2013). *Performativity and event in 1960s Japan: City, body, memory.* Houndmills/Basingstoke/Hampshire/New York: Palgrave Macmillan.

Eckersall, P. (2015). Towards a dramaturgy of robots and object-figures. *TDR/The Drama Review, 59*(3), 123–131. doi:10.1162/DRAM_a_00474.

Feenberg, A. (2010). *Between reason and experience essays in technology and modernity.* Cambridge, MA: MIT Press.

Fensham, R. (2009). *To watch theatre: Essays on genre and corporeality.* Bruxelles/New York: P.I.E. Peter Lang.

Fischer-Lichte, E. (2008). *The transformative power of performance: A new aesthetics* (S. I. Jain, Trans.). New York: Routledge.

Fischer-Lichte, E. (2014). *The Routledge introduction to theatre and performance studies.* M. Arjomand & R. Mosse (Eds.) (M. Arjomand, Trans.). London/New York: Routledge.

Ford, M. (2015). *Rise of the robots: Technology and the threat of a jobless future.* New York: Basic Books.

Freshwater, H. (2009). *Theatre & audience*. Houndmills/Basingstoke/ Hampshire/New York: Palgrave Macmillan.

Funabiki, T. (2010). *Nihonjinron saikō* [A Reexamination of Nihonjinron]. Tokyo: Kōdansha.

Geraci, R. M. (2010). *Apocalyptic AI visions of heaven in robotics, artificial intelligence, and virtual reality*. Oxford/New York: Oxford University Press.

Giannachi, G. (2004). *Virtual theatres: An introduction*. London/New York: Routledge.

Giannachi, G. (2007). *The politics of new media theatre: Life TM*. London/New York: Routledge.

Goldblatt, D. (2006). *Art and ventriloquism*. London/New York: Routledge.

Grehan, H. (2009). *Performance, ethics and spectatorship in a global age*. Basingstoke/New York: Palgrave Macmillan.

Grehan, H. (2015). Actors, spectators, and 'vibrant' objects: Kris Verdonck's ACTOR #1. *TDR: The Drama Review, 59*(3), 132–139. doi:10.1162/ DRAM_a_00475.

Guillot, A., & Meyer, J. A. (2010). *How to catch a robot rat: When biology inspires innovation* (S. Emanuel, Trans.). Cambridge, MA: MIT Press.

Harootunian, H. D. (2000a). *History's disquiet modernity, cultural practice, and the question of everyday life*. New York: Columbia University Press.

Harootunian, H. D. (2000b). *Overcome by modernity history, culture, and community in interwar Japan*. Princeton: Princeton University Press.

Heim, C. (2016). *Audience as performer: The changing role of theatre audiences in the twenty-first century*. London/New York: Routledge.

Hibino, K. (2012). Oscillating between Fakery and Authenticity: Hirata Oriza's Android Theatre. *Comparative Theatre Review, 11*(1), 30–42. doi:10.7141/ctr. 11.30.

Hirakawa, S. (2006). *Wakon yōsai no keifu: Uchi to Soto kara no Meiji nihon, jō* [Genealogy of Japanese spirituality and Western Technology: Meiji Japan inside and outside, Part 1]. Tokyo: Heibonsha.

Hornyak, T. N. (2006). *Loving the machine: The art and science of Japanese robots* (1st ed.). Tokyo/New York: Kodansha International.

Hotta, J. (2008). *Hito to robotto no himitu* [The secret concerning robots and humans]. Tokyo: Kōbunsha.

Huyssen, A. (1986). *After the great divide: Modernism, mass culture, postmodernism*. Bloomington: Indiana University Press.

Ishiguro, H. (2007). *Andoroido saiensu: Ningen wo sirutame no robotto kenkyū* [Android science: A study to learn what the human is]. Tokyo: Mainichi Komyunikēshonzu.

Jackson, S. (2011). *Social works: Performing art, supporting publics*. New York: Routledge.

Kac, E. (2005). *Telepresence & bio art: Networking humans, rabbits & robots*. Ann Arbor: University of Michigan Press.

Kang, M. (2011). *Sublime dreams of living machines the automaton in the European imagination*. Cambridge, MA/London: Harvard University Press.

Kaplan, F. (2011). *Robotto wa tomodachi ni nareruka: nihonnjin to kikai no fusigi na kankei* [Can a robot be our friend: The mysterious relationship between the Japanese and robots]. Translated from French to Japanese by Kenji Nishi. [*Les Machines Apprivoisées: Comprendre les Robots de Loisir*]. Tokyo: NTT Shuppan.

Kennedy, D. (2009). *The spectator and the spectacle: Audiences in modernity and postmodernity*. Cambridge/New York: Cambridge University Press.

Kishi, N. (2011). *Robotto ga nihon o suku'u* [Robots will save Japan]. Tokyo: Bungē shunjū.

Klich, R., & Scheer, E. (2012). *Multimedia performance*. Houndmills/Basingstoke/Hampshire/New York: Palgrave Macmillan.

Kubo, A. (2015). *Robotto no jinruigaku: Nijūseiki no kikai to ningen* [Anthropology of the robot: The machine and the human in the 20th century]. Tokyo: Sekai Sisōsha.

LaMarre, T. (2009). *The anime machine: A media theory of animation*. Minneapolis: University of Minnesota Press.

Levy, D. N. L. (2007). *Love and sex with robots: The evolution of human-robot relations*. New York: HarperCollins.

Lin, P., Abney, K., & Bekey, G. A. (2012). *Robot ethics: The ethical and social implications of robotics*. Cambridge, MA: The MIT Press.

Loxley, J. (2007). *Performativity*. London/New York: Routledge.

MacDorman, K. F., Vasudevan, S. K., & Ho, C.-C. (2009). Does Japan really have robot mania? Comparing attitudes by implicit and explicit measures. *AI & Society, 23*(4), 485–510. doi:10.1007/s00146-008-0181-2.

Matsubara, H. (1999). *Tetsuwan Atomu wa jitsugen dekiruka: Robo kappu ga hiraku mirai* [Would we be able to actualise astro boy?: The future created by RoboCup]. Tokyo: Kwade Shobō.

McConachie, B. A. (2008). *Engaging audiences: A cognitive approach to spectating in the theatre*. New York/Basingstoke: Palgrave Macmillan.

Meadows, M. S. (2011). *We, robot: Skywalker's hand, blade runners, Iron Man, slutbots, and how fiction became fact*. Guilford: Lyons Press.

Menzel, P., & D'Aluisio, F. (2000). *Robo sapiens: Evolution of a new species*. Cambridge, MA: MIT Press.

Miyoshi, M., & Harootunian, H. D. (1989). Introduction. In M. Miyoshi & H. D. Harootunian (Eds.), *Postmodernism and Japan* (pp. vii–xix). Durham: Duke University Press.

Mori, M. [Masahiro]. (2012). The uncanny valley [From the field] (K. F. MacDorman & N. Kageki, Trans.). *IEEE Robotics Automation Magazine, 19*(2): 98–100. doi:10.1109/MRA.2012.2192811.

Mori, M. [Masahiro]. (2014). *Robotto kōgaku to ningen: Mirai no tameno robotto kōgaku* [Robotics and the human: Robotics for the future]. Tokyo: Ōmusha.

Mori, M. (2002). The structure of theater: A Japanese view of theatricality. *Substance, 31*(2), 73–93. doi:10.1353/sub.2002.0033.

Morley, D., & Robins, K. (1995). *Space of identity: Global media, electronic landscapes and cultural boundaries.* London/New York: Routledge.

Morse, M. (2016). *Soft is fast: Simone Forti in the 1960s and after.* Cambridge, MA: MIT Press.

Murata, J. (2003). Creativity of technology: An origin of modernity?. In T. J. Misa, P. Brey, & A. Feenberg (Eds.), *Modernity and technology* (pp. 151–177). Cambridge, MA: MIT Press.

Najita, T. (1989). On culture and technology in postmodernism and Japan. In M. Miyoshi & H. D. Harootunian (Eds.), *Postmodernism and Japan* (pp. 3–20). Durham: Duke University Press.

Nakamura, M. (2007). Horror and machines in Prewar Japan: The mechanical Uncanny in Yumeno Kyūsaku's Dogura Magura. In C. Bolton, I. Csicsery-Ronay, & T. Tatsumi (Eds.), *Robot ghosts and wired dreams Japanese science fiction from origins to anime* (C. Bolton, Trans.) (pp. 3–26). Minneapolis: University of Minnesota Press.

Nakayama, S. (2006). *Robotto ga nihon o suku'u* [Robots will save Japan]. Tokyo: Toyō Keizai Shinpōsha.

Nelson, V. (2001). *The secret life of puppets.* Cambridge, MA: Harvard University Press.

Nishida, K. (2007). *Essensharu Nishida, soku no kan, Nishida Kitaro kīwādo ronshū* [Essential Nishida: An issue on instantaneity: Essays that contain Kitaro Nishida's Keywords]. Tokyo: Shoshi Shinsui.

Numata, H. (2001). Robokappu no sozōryoku [RoboCup imagination]. In H. Matsubara, I. Takeuchi, & H. Numata (Eds.), *Robotto no jōhōgaku: 2050 nen wārudo kappu ni katsu* [Robot information study: Winning the World Cup in 2050] (pp. 117–155). Tokyo: NTT Shuppansha.

Oddey, A., & White, C. (2009). *Modes of spectating.* Bristol/Chicago: Intellect.

OED (Online). (2000). Oxford: Oxford University Press.

Okada, M. (2012). *Yowai robotto* [Weak robot]. Tokyo: Igaku Shoin.

Okuno, T. (2002). *Ningen dōbutsu kikai: Tekuno animizumu* [Human, animal, machine: Techno animism]. Tokyo: Kadokawa shoten.

Orikuchi, S. (1991). *Nihon geinōshi rokkō* [The history of Japanese entertainment: Six lectures]. Tokyo: Kōdansha.

Parker-Starbuck, J. (2011). *Cyborg theatre: Corporeal/technological intersections in multimedia performance.* Houndmills/Basingstoke/Hampshire/New York: Palgrave Macmillan.

Parker-Starbuck, J. (2015). Cyborg. Returns: Always-already subject technologies. In S. Bay-Cheng, J. Parker-Starbuck, & D. Z. Saltz (Eds.), *Performance and media: Taxonomies for a changing field* (pp. 65–92). Ann Arbor: University of Michigan Press.

Perkowitz, S. (2004). *Digital people: From bionic humans to androids.* Washington, DC: Joseph Henry Press.

Posner, D. N. (2014). Contemporary investigations and hybridizations. In D. N. Posner, C. Orenstein, & J. Bell (Eds.), *The Routledge companion to puppetry and material performance* (pp. 225–227). London/New York: Routledge.

Poulton, C. (2014). From puppet to robot: Technology and the human in Japanese theatre. In D. N. Posner, C. Orenstein, & J. Bell (Eds.), *The Routledge companion to puppetry and material performance* (pp. 280–293). London/New York: Routledge.

Power, C. (2008). *Presence in play a critique of theories of presence in the theatre.* Amsterdam/New York: Rodopi.

Radbourne, J., Glow, H., Johanson, K., Thomas, E., & Marshall, M. (Eds.). (2013). *The audience experience: A critical analysis of audiences in the performing arts.* Bristol/Chicago: Intellect.

Rancière, J. (2009). *The emancipated spectator* (G. Elliott, Trans.). London: Verso.

Raz, J. (1983). *Audience and actors: A study of their interaction in the Japanese traditional theatre.* Leiden: E.J. Brill.

Reilly, K. (2011). *Automata and mimesis on the stage of theatre history.* Houndmills/Basingstoke/Hampshire/New York: Palgrave Macmillan.

Reinelt, J. G. (2002). The politics of discourse: Performativity meets theatricality. *SubStance, 31*(2), 201–215. doi:10.1353/sub.2002.0037.

Robertson, J. (2007). Robo sapiens Japanicus: Humanoid robots and the posthuman family. *Critical Asian Studies, 39*(3), 369–398. doi:10.1080/14672710701527378.

Robertson, J. (2010). Gendering humanoid robots: Robo-sexism in Japan. *Body & Society, 16*(2), 1–36. doi:10.1177/1357034X10364767.

Robertson, J. (2014). Human rights vs. robot rights: Forecasts from Japan. *Critical Asian Studies, 46*(4), 571–598. doi:10.1080/14672715.2014.960707.

Šabanović, S. (2014). Inventing Japan's 'robotics culture': The repeated assembly of science, technology, and culture in social robotics. *Social Studies of Science, 44*(3), 342–367. doi:10.1177/0306312713509704.

Salter, C. (2010). *Entangled: Technology and the transformation of performance.* Cambridge, MA: MIT Press.

Sandry, E. (2015). *Robots and communication.* New York: Palgrave Macmillan.

Schechner, R. (2013). *Performance studies : An introduction* (3rd ed.). New York: Routledge.

Scheer, E. (2015). Robotics as new media dramaturgy: The case of the sleepy robot. *TDR: The Drama Review, 59*(3), 140–149. doi:10.1162/DRAM_a_00476.

Schodt, F. L. (1988). *Inside the robot kingdom: Japan, mechatronics, and the coming robotopia.* Tokyo/New York: Kodansha International.

Sena, H. (2001). *Robotto Seiki*. Tokyo: Bungei Shunjū.

Sena, H. (Ed.). (2004). *Robotto Opera (Robot Opera: An anthology of robot fiction and robot culture, original English title)*. Tokyo: Kōbunsha.

Sena, H. (2008). *Sena Hideaki robottogaku ronshū* [Hideaki Sena robot study essay collection]. Tokyo: Keisō Shobō.

Shershow, S. C. (1995). *Puppets and 'popular' culture*. Ithaca: Cornell University Press.

Silverberg, M. R. (2006). *Erotic grotesque nonsense the mass culture of Japanese modern times*. Berkeley: University of California Press.

Smith, M. (Ed.). (2005). *Stelarc: The monograph*. Cambridge, MA/London: MIT Press.

Starrs, R. (2011). *Modernism and Japanese culture*. Houndmills/ Basingstoke/New York: Palgrave Macmillan.

Sugano, S. (2011). *Hito ga mita yume, robotto no kita michi: Girisha shinwa kara Atomu, soshite* [Mankind's Dream, The Historical Path: The Robot From (ancient) Greece to Astro Boy]. Tokyo: JIPM Sorūshon.

Suzuki, K. (2007b). Karakuri ga hagukunda nihon no robotto kan. In K. Suzuki (Ed.), *Dai robotto haku: Karakuri kara anime, saishin robotto made* [The Great Robot Exhibition: From karakuri, to anime, to the latest robots] (pp. 44–49). Exh. cat. Tokyo: Yumiuri Shimbun.

Suzumori, K. (2012). *Robotto wa naze ikimono ni niteshimaunoka: Kōgaku ni tachihadakaru kyūkyoku no rikigaku kōzō* [How would a robot become similar to a living creature: The ultimate principle of dynamics before engineering]. Tokyo: Kōdansha.

Takano, Y. (2008). *Shūdan shugi toiu sakkaku: Nihonjinron no omoichigai to sono yurai* ['Groupism' as Illusion: The Misunderstanding Concerning Nihonjinron and its Origin]. Tokyo: Shinchōsha.

Takeuchi, Y. (2007). Ējent medieiteddo intarakushon [Agent-mediated interaction]. In S. Yamada (Ed.), *Hito to robotto no aida wo dezainsuru* [Designing that which is in-between Humans and Robots] (pp. 259–288). Tokyo: Tokyo Denki Daigaku Shuppankyoku.

The International Federation of Robotics (IFR). (2015). Industrial robots: IFR International Federation of Robotics. http://www.ifr.org/industrial-robots/. Accessed on 10-05-2015.

Tillis, S. (1992). *Toward an aesthetics of the puppet: Puppetry as a theatrical art*. New York: Greenwood Press.

Toida, M. (1994). *Engi* [Acting]. Tokyo: Kinokuniya Shoten.

Turkle, S. (2007). *Evocative objects things we think with*. Cambridge, MA: MIT Press.

Turkle, S. (2011). *Alone together: Why we expect more from technology and less from each other*. New York: Basic Books.

Turner, V. (1967). *The forest of symbols: Aspects of Ndembu ritual.* Ithaca: Cornell University Press.

Umetani, Y. (2005). *Robotto no kenkyūsha wa gendai no karakurisi ka?* [Are robot researchers contemporary Karakuri Craftsmen?]. Tokyo: Ōmusha.

Wallach, W., & Allen, C. (2009). *Moral machines: Teaching robots right from wrong.* Oxford/New York: Oxford University Press.

Walton, K. L. (1990). *Mimesis as make-believe: On the foundations of the representational arts.* Cambridge, MA: Harvard University Press.

White, G. (2013). *Audience participation in theatre: Aesthetics of the invitation.* Houndmills/Basingstoke/Hampshire/New York: Palgrave Macmillan.

Wood, G. (2003). *Edison's Eve: A magical history of the quest for mechanical life.* New York: Anchor Books.

Yablonsky, L. (1972). *Robopaths.* Indianapolis: Bobbs-Merrill.

Yamaguchi, M. (2002). The ludic relationship between man and machine in Tokugawa Japan. In J. Hendry & M. Raveri (Eds.), *Japan at play: The ludic and logic of power* (pp. 72–83). London/New York: Routledge.

Yonezawa, Y. (Ed.). (2002). *Robotto manga wa jitsugen suruka* [Could Robot Manga be Actualised]. Tokyo: Jitsugyō no Nihonsha.

Yoshimi, S. (1999). "Made in Japan": The cultural politics of 'home electrification' in Postwar Japan. *Media, Culture & Society, 21*(2), 149–171. doi:10.1177/016344399021002002.

CHAPTER 2

Robotics and Representation

In this chapter, I explain the nature of the affective power that, for Japanese roboticists and the Japanese viewers, seems to characterise the appearance of next-generation humanoids. I discuss the relation of manga/anime to roboticists' conceptions of the humanoid and consider particularly Japanese approaches to the idea of 'presence' through concepts such as aura and icon. I also look at the idea of presence in the context of theatre and performance studies.

At the beginning of the twenty-first century, several Japanese mega-corporations unveiled significant prototypes in their humanoid robotics research. For example, car company Honda presented ASIMO (Advanced Step in Innovative Mobility) in 2000 (Fig. 2.1). According to Masato Hirose and Ken'ich Ogawa, the company's researchers responsible for the development of the bipedal robot, ASIMO means '"asita-no" mobility' ('*asita*' mean 'tomorrow' in Japanese, hence, mobility in/for the future) (2007, 14). This boy-like robot (1.2 metres in height) is able to walk and run as well as dance with what might be seen as gestural expressiveness.[1] Toyota also debuted a 1.5 m-high humanoid trumpet player as part of CONCERO, a group of musician robots, in 2005. With highly developed posture sensors, this anthropomorphic bipedal robot can play a trumpet with its fingers while blowing air into the instrument, while it is walking. Consumer electronics firm Sony presented a compact humanoid of a half-metre in size with a silver magnesium alloy body, SDR-3X (where SDR stands for Sony Dream Robot) in 2000, followed by QRIO (Quest for Curiosity) in 2003, which are much faster and more agile in their

© The Author(s) 2017
Y. Sone, *Japanese Robot Culture*,
DOI 10.1057/978-1-137-52527-7_2

Fig. 2.1   ASIMO, Honda's humanoid robot (2008 © Kyodo News. Courtesy of Kyodo News International)

movements than are ASIMO and CONCERO when they are dancing. QRIO is equipped with face and voice recognition capabilities and communicates with humans 'via gestures, lights in its eyes and chest and a squeaky, high-pitched voice' (Hornyak 2006, 112) (Fig. 2.2). Also, construction and steel manufacturing firm Kawada Industries developed and presented approximately 1.5 m-high HRP-2 Promet in 2002, taking part in the Humanoid Robotics Project (1998–2003), a large-scale research project funded by the Ministry of Economy, Trade and Industry (METI) through its New Energy and Industrial Technology Development Organization (NEDO).[2] Unlike

Fig. 2.2 Sony's 'Aibo' and 'Qrio', TOKYO, Japan—Sony demonstrates its robots Aibo (*right*) and Qrio (2004 © Kyodo News. Courtesy of Kyodo News International)

the other humanoids above, Promet is designed for outdoor use. This bipedal robot is capable of lying down and rising from a prone position, and it is the first humanoid of an adult human's size to be able to do so.

The explanations by these corporations concerning their development of these next-generation robots, as evident in their promotional materials, reveal their ideas for the future of human–robot co-existence and co-habitation as preconceived. Honda states that at the beginning of the company's robot project in 1986 the aim was to develop a robot as a valuable research platform concerning machine mobility, highlighting the notions of co-existence and incorporation in society. For the ASIMO project, it was assumed that the robot would have to work in human living spaces (Honda Motor Company 2015a).[3] Similarly, Soya Takagi, who was responsible for Toyota's partner robot project, explains that the project's aim was 'the creation of [a] robot that can use tools, assist people, and live in harmony' with humans (Toyota Motor Corporation 2015, original English). In the case of Sony, current technological limitations concerning humanoid development led to the creation of entertainment robots such as QRIO as well as AIBO (a robot dog). (I will discuss AIBO in Chap. 8.) According to Toshitada Doi, the scientist and electronics engineer who was responsible for the development of Sony's robot project, while it was impossible to build a completely autonomous interactive robot with current technologies, it was deemed feasible to create an entertainment robot whose key task would be to appeal to a user (2012, 1000). Nevertheless, Doi had a vision for a new kind of robot industry, facilitating robotics that would have direct engagement with humans (2012). Masahiro Fujita, one of the key members in Sony's robot project, argues for the importance of entertainment robot research in terms of human communication if robots are to coexist with humans (interviewed by Yonezawa 2002, 318). Kawada Industries continued to develop the bipedal humanoid alongside static humanoids for factory assembly lines. HRP-3, developed in 2007, has articulated hands that can use electric building tools, while HRP-4, built in 2010, is a female-looking humanoid with a much lighter and slimmer body. Kawada Industries' motivation for their robot projects is more pragmatic in accordance with the main objective of the Humanoid Robotics Project, which is to demonstrate the possibilities for realising and implementing real working humanoids in order to expand the current markets for robot use (The Humanoid Research Group of National Institute of Advanced Industrial Science and Technology 2015).

In the early twenty-first century, these prototype robots were promoted widely in Japanese mass media as interactive robots intended to be used in close proximity to humans. The development of these prototypes prompted a discussion of whether or not it is possible to create a robot just like Astro Boy. According to Osamu Tezuka's seminal comic and cartoon originally created in the early 1950s, the friendly child robot Astro Boy was born in 2003, a boom time for the development of next-generation robots. This date may well have seemed symbolic for the generation of roboticists who grew up reading *Astro Boy* comics and watching the television cartoon, in circulation in Japanese popular media for decades, well after the 1950s. I suggest in this chapter that such ideas may have influenced these roboticists, offering goals and ideal forms for their productions. While roboticists have expressed themselves in the technical language used in engineering, cognitive science, and robotics for discussions within these fields, references to Japanese culture as well as to popular Japanese robot manga/anime have often appeared in their writings for general publications. These are book publications that are designed for general educational purposes and for the promotion of their ideas in interview, conversational, or autobiographical formats.[4] The discussions in these popular book publications provide the usual narrative concerning the popularity of humanoids in Japan, which consistently relates the development of humanoids to manga/anime as a kind of cultural backdrop and reference point. Media theorist Nobuhiko Baba states that the research motivations for Japanese robotics are to a large extent sustained by images drawn from science fiction narratives, animations, and films (2004, 11). The 'newsworthiness' and utility of the next-generation robots are enhanced by a context of readiness, emblematised by the popularity of Astro Boy in Japan (Yamada 2013, 52). Such popular-culture references are used by roboticists in their writings for a general audience as part of their justifications for their humanoid research. In short, the theme of the humanoid robot triggers certain narratives already in social circulation, and these narratives give rise to a sense of 'eventful-ness', according to science fiction writer and expert on robot culture Hideaki Sena (2008, 384 and 485).

In the Japanese context, the superhero or friendly humanoid robot is a concept in the realm of what philosopher Charles Taylor calls the 'social imaginary', a concept I raised in the introduction to this book (2004, 23). For Taylor, the social imaginary provides the background to 'the deeper normative notions and images' that people use to imagine their existence in society and guides behaviours and attitudes, incorporating 'some sense of how we all fit together in carrying out the common practice' (2004, 23–4). The popular image of the humanoid has become naturalised among the post-war generations of Japanese. The pervasiveness of popular robot manga/anime is such that if ordinary Japanese people were asked what robots they are familiar with, they would most likely point to those from manga/anime (Yamato 2006, 4). The general view of the robot in Japan is often driven by these popular visual images of the robot, interacting with other forms of robots with which people may be familiar, such as the industrial robot or hobby robot.

While roboticists and cultural theorists do not consciously discuss them, such representational and contextual assumptions inform the designs of the next-generation robots of human-like appearance. For example, it is not surprising that promotional photos of actual humanoid robots can look like robots as they are depicted in manga/anime. In this chapter, I shine light on how the ideas of the humanoid, which are drawn from manga/anime, interact with the concepts and designs of the next-generation humanoids. The proponents of the humanoid, consciously or unconsciously, draw upon the idea of the robot in popular media and social fantasy when designing and imaging humanoid robotic technology (Baba 2004, 11). Cultural narratives around robot manga/anime, such as that of Astro Boy as the saviour of Japan, upon which I will elaborate, interact with images of, and events featuring, anthropomorphic robots, to facilitate a sense of the robots' 'presence'. Before discussing this presence in these specific encounters in the following chapters, this chapter discusses how the image of Astro Boy, to select one prominent model, can induce a sense of presence for viewers concerning next-generation robots such as ASIMO. I first examine writings by Japanese roboticists and cultural theorists who argue for humanoid research, and discuss popular-culture representations of the robot in Japan.

Japanese Humanoid Research

The arguments for the humanoid usually take two directions: research and application. Japanese research into bipedal machines began in the 1960s, by pioneers of Japanese robotics such as Masahiro Mori and Ichiro Kato, as a way to apply and test system control theories, the kinematics and dynamics of walking machinery, or towards the application of biomechanics (Nagata 2005, 147). For Atsuo Takanishi, one of the key successors of Kato's humanoid research at Waseda University, the humanoid as a model of the human could be used to collect analogue data on human sensing (interviewed by Yonezawa 2002, 142–3). Thus, as a research platform, a humanoid is useful, according to Sena, in developing related engineering technologies and software in mobility and communication as well as in conducting studies in relation to brain science, cognitive science, or sports science (2001). Roboticists Minoru Asada and Yasuo Kuniyoshi, with Hiroshi Ishiguro, have coined the term 'cognitive developmental robotics' to explain their research as a new area of integrative study for robotics, in which humanoids are used to study how machines can learn perception and motor skills in relation to the development of human intelligence (Asada 2010, 91; A. Kubo 2015, 180–1).

Mori and Kato thus laid the foundations for Japanese humanoid research. Mori advocated the notion of the 'soft machine' in the late 1970s as opposed to the 'hard machine': unlike the latter, which is understood as the industrial robot that only pursues productivity and efficiency, the soft machine is a concept that describes a robot that can facilitate harmony between humans and machines. In the same period, Kato explored the notion of the 'artificial hand' as opposed to the 'machine hand'. While the latter was a machine that sought to emulate the efficiency and dexterity of the human hand, Kato's concept was more concerned with uniquely human gestures, such as shaking someone's hand. Kato extended his investigations towards the artificial leg, which led to his well-known research on the bipedal humanoid. In the same way that the term 'my car', an English phrase used in Japanese speech, refers to a car as a common household item, Kato's concept of 'my robot' in the 1980s suggests a future when robots become commonly used at home and the service robot industry flourishes. Kubo points out that while Mori and Kato did not advocate for the development of humanoid robots as such, the common theme between them is the notion of a machine that is flexible and able to adapt itself to human needs (2015, 156).

It is generally regarded by the proponents of the next-generation anthropomorphic robots that for stable interaction or establishing a connection between robots and users, a certain degree of anthropomorphism is encouraged as it is thought to make the robot seem more familiar. An anthropomorphic robot is regarded as able to induce a temporary sense that the robot is intelligent and is capable of emotional response (Ishiguro 2009, 220). In other words, a combination of 'external' (human-like appearance) and 'internal' (human-like behaviour or action) anthropomorphism establish 'agency' in a robot (Nishida 2005, 92).

According to Sena, the proponents of humanoid research and development put forward two main arguments: it is better for a robot, if it is to be used in a house, to have a human-sized and -shaped body as the human living environment is designed for human bodies, and the humanoid is economical because of its general versatility (2004, 211, 2008, 97). Roboticist Hitoshi Matsubara argues that it would be better for elderly people who are in need of care to be able to communicate with assistive machines, and a humanoid would be advantageous in household use because of its mobility in a house, and its human-like communication capabilities (1999, 22–5, 2001, 27–8). Indeed, at the start of Honda's humanoid research in the mid-1980s, the aim was 'to develop a more viable mobility that would allow robots to help and live in harmony with people' in places where the robot has to move between objects and climb up and down steps, and for that reason, 'it had to have two legs, just like a person' (M. Hirose and

Ogawa 2007, original English). The proponents of the humanoid suggest that humanoids can also be used as entertainment robots, as discussed by Fujita in relation to Sony's project (interviewed by Yonezawa 2002, 318).

However, not all Japanese roboticists are advocates for the development of humanoids: humanoid research has its critics, and criticism against humanoid research intensified after the Fukushima nuclear disaster in 2011. Despite their faith in Japan as the world's leading industrial robot manufacturer, according to media reports, Japanese roboticists were humiliated by the fact that American rescue robots (Packbot 510, designed by iRobot) were used during the crisis at the Fukushima nuclear plant (Ishihara 2011). Despite the fact that there was a decade-long national project in the 1980s—the RCW (the robots for critical work) that was developing robots which could perform in hazardous environments—Japanese technology was inadequate for the work that needed to be done at the Fukushima site.[5]

Shigeo Hirose states that arguments for utility and social function in fact constitute a smokescreen for researchers who simply want to build humanoids (2011, 138–9). Hiroshi Kobayashi similarly argues for practical and pragmatic robotic research, criticising the unclear goals of Japanese bipedal robot research (2006, 49). It has been observed that in humanoid research, there is not enough thought about the ordinary people who would have to live with humanoids (S. Hirose 2011, 139). Yukio Honda also points out the lack of strategies regarding product development for the global marketplace because of the gap between the university, the realm of concepts and ideas, and the industry (2014, 118–9). The common points in these criticisms are that the arguments for Japanese humanoid research are circular, the relation of the aims of research to the applications of the research is unclear, and proponents cannot explain why human-like robots are actually necessary; rather, researchers are just pursuing a dream of a thinking bipedal robot that is as close as possible to human in every way (Doi 2012, 68–9). Indeed, Baba questions if researchers are really just waiting for a future when humans and robots coexist, as is portrayed in the manga/anime *Astro Boy* (Baba 2004, 10). Remarks such as Baba's could not show the pervasive influence of Japanese manga/anime, the seedbed for the roboticists of the next-generation humanoids, any more starkly.

Despite criticisms concerning their actual utility, it is certainly the case that bipedal machines are always popular—that a robot that is able to walk excites an audience of ordinary Japanese people (Sena 2001, 44). As a result of the portrayal of humanoids in the popular media, humanoid research has gained a high profile in the public's awareness of contemporary science, even though the actual number of Japanese roboticists who dedicate them-selves for humanoid research is less than 10% of all roboticists (Kajita 2008, 50). In the next section of this chapter, I will discuss specific encounters between roboticists in humanoid research and examples of manga/anime to consider what they are seeing in these examples and incorporating into their designs, and how the humanoids are imaged and presented.

Japanese Robotics and Manga/Anime Culture

It is known that Masato Hirose, Honda's roboticist responsible for ASIMO, stated that the question that initially spurred him on to develop ASIMO was put forward by his superior: 'Do you [Hirose] want to try making a robot like Astro Boy?' (2006, 116). This story of the genesis of Honda's ASIMO is a good example of the productive entanglement of robotics and popular imagination concerning the robot in Japan. As I have detailed in the beginning of this chapter, next-generation robotics has often been discussed by cultural theorists in Japan in relation to robot manga/anime. Manga critic Yoshihiro Yonezawa's book (2002)—its title can be translated as *Could Robot Manga Be Actualised?*—consists of chapters that discuss the next-generation robots, juxtaposing robot manga comics and comments by Japanese roboticists within an interview format. Non-fiction writer Junji Hotta (2008) also deploys the formula of interviewing roboticists with manga/anime references, while writer Rikao Yanagida and toy robot crea-tor Tomotaka Takahashi (2011) record their conversations on robot manga/anime. In general, publishers are aware that linking Japanese robot manga/anime with humanoid research would appeal to the general Japanese audience.

These writers discuss classic robot manga/anime in which humanoids are the main characters—such as *Astro Boy* (1951); *Iron Man No. 28* (1956), which is about a remote-controlled giant super robot; *Mazinger Z* (1972), whose eponymous robot is a ride-able giant combat machine; *Mobile Suit Gundam* (1979), which is about a giant exoskeletal combat system; and *Doraemon* (1970), whose protagonist is a friendly, anthropomorphised cat-like robot. Why are these examples of robot manga/anime associated with next-generation robots? This question can be answered in terms of the possible influences of these highly visual stories on post-war generations of the roboticists (and writers), as has been noted more generally, and in terms of certain representational resemblances.

Astro Boy's friendly, boy-like appearance can be detected in ASIMO and Sony's QRIO. Because HRP-2's appearance was designed by Yutaka Izubuchi, a designer for robot costume and characters in animation, including Gundam, the robot is reminiscent of one of those characters. Both Toyota's i-foot, a two-legged walking machine with a halved eggshell-like cockpit, and Tmsuk's T-52 Enryū, a rescue robot with two gigantic arms controlled by an operator in the cockpit on a locomotive base, resemble Mazinger Z's cockpit. It is possible to see a thematic link between roboticist Michio Okada's pursuit of imperfect robots that draw children's attention and the friendly Draemon character, who is a defective service and partner robot working in a domestic situation. As for technology-based thematic links, the idea of human–machine symbiosis can be seen in roboticist Yoshiyuki Sankai's HAL (Hybrid Assistive Limb), a wearable robot suit that utilises the nerve signals of an elderly or disabled patient to assist in walking and carrying things. Sankai acknowledges that he has been influenced by *Cyborg 009* (1968), a story about cyborg rangers who fight against villains (2006, 112).

The Japanese roboticists who have publicly acknowledged the influences of Japanese robot manga/anime on their work include Minoru Asada (2010, 138), Takayuki Furuta (2006, 12), Yoshiyuki Sankai (2006, 112), Atsuo Takanishi (interviewed by Yonezawa 2002, 137), Shigeki Sugano (2011, 4), Jun'ichi Takeno (2011, i), and Hitoshi Matsubara (1999, 7), to name a few. Takanishi and Yonezawa are in agreement that their generation of Japanese who were born between the mid-1950s and the mid-1960s were exposed to robot manga/anime on a daily basis while growing up, and Takanishi admits that from childhood he had a dream to build a robot (2002, 137). Most of the introductory books on the next-generation robots by these roboticists focus on the post-war Japanese robot manga/anime up to *Mobile Suit Gundam* of the late 1970s and the 1980s, a time when the roboticists working on next-generation robots would have been teenagers or young adults. Their books are less inclined to discuss notable works of the 1980s and 1990s, such as *Ghost in the Shell* (1989) and *Neon Genesis Evangelion* (1995).

These connections between robot manga/anime images and next-generation robots seem to demonstrate the point made by Sena that '[i]mages are being passed back and forth between fiction and real-life science, and these two realms are closely interconnected' (quoted by Hornyak 2006, 56). In general, both Japanese robot manga/anime and the next-generation robots suggest and promote a future when robots are common in everyday life. It is this relay connecting Japanese fantasy, representation, and robot that I am interested in here.

It is, of course, also the case that Western popular culture has influenced Western roboticists' thinking. Cynthia L. Breazeal, a well-known roboticist who has created a social robot, Kismet, acknowledges her fascination for the friendly robot characters of *Star Wars*, R2-D2, and C-3PO. She likes them, in contrast to the 'cool' and 'eerie' computer HAL in *2001: A Space Odyssey* (2002, xi). Western narratives do feature some friendly robots, like the animated film *Wall-E* (2008) or *Big Hero 6* (2014), or *Transformers* (2007), which is about alien robots who fight for Earth. However, like the HAL computer, the robotic characters of Western science fiction are not always friendly and are more often threatening, in contrast to their Japanese counterparts.

Indeed, in the Western modern imaginary, robots are often seen as not only imperfect but also sinister replacements for humans, such as the character of robot Maria in the film *Metropolis* (1926). Even if they are not sinister, they are artificial beings who want to be human and occupy a liminal narrative space, such as Data in *Star Trek: The Next Generation* (1987) and as we see in *Bicentennial Man* (1999) and *A.I.* (2001), which feature a *Pinocchio* or *Frankenstein* narrative. Alternatively, robots are portrayed as machines that become enemies of humanity, such as Gort in *The Day The Earth Stood Still* (1951) or the Daleks in the *Doctor Who* television series (which aired on British television starting in 1963). Where the robot is a hero it is also, at the same time, a potential threat, as in the films *Terminator* (1984) and *RoboCop* (1987). Science fiction writer Isaac Asimov refers to the fear of the robot and the narrative of a vendetta between a robot and its creator scientist as the 'Frankenstein Complex' (Sena 2004, 11). Asimov's well-known three commandments for robots in *I, Robot*, published in 1950, reflect this culturally embedded fear of robots.[6] The robot servants of *I, Robot* (2004) and the Swedish SF television drama *Real Humans* (first season in 2012) rebel against humans, revisiting the *R.U.R.* theme of dehumanisation related to its narrative of robots exploited for their labour. Literature and film studies scholar Despina Kakoudaki argues that Čapek's *R.U.R.* reflects conditions for workers under capitalism; it is an allegorical tale about 'human otherness, racial difference' and 'slavery' (2014, 142 and 144). For religious scholar Robert Geraci, Western science fiction, reflecting a dualistic view of the apocalyptic, tends to regard intelligent machines as the embodiment of possibilities for both human enslavement and happiness, and, hence, the symbol of both 'damnation and salvation – which leads to fear and fascination intertwined' (2010, 50).

In contrast, the basic principle for robot characters in Japanese comics and animation is that a robot protagonist, as a saviour, fights for the good side—that is, it fights for Japan against villains that include evil and/or enemy robots or is a friendly offbeat character who helps humans. Recent Japanese filmmakers have explored human–robot relations in ways that reflect social reality, as in *Hinokio* (2004), in which the main character is a *hikikomori*, a reclusive schoolboy who uses a robot avatar to attend school in his place, and in *Robo-G* (2012), a comedy in which electronics engineers, out of desperation to present a new, advanced robot, present a fake robot that is in fact operated by a retired old man inside it.

Of course, in Japanese films and animation there are works that deal with dark themes, as in the film *Tetsuo the Iron Man* (1989), in which metallic machine parts grow from the body of the main character, and the animation *Roujin Z* (1991), in which the nuclear-power-driven robotic bed Z-001, designed for old-age care, is in fact an experimental weapon. These works attract non-mainstream audiences. Using the phrase 'Frankenstein syndrome', scholar of Japanese literature and popular culture Sharalyn Orbaugh observes 'the tendency to explore monstrous subjectivities from a sympathetic, interiorized point of view' in Japanese manga/animation (2005, 62). Alexandra Munroe, acclaimed curator and historian of modern and contemporary Asian art, critically observes that, '[f]ocused often on apocalyptic imagery, with frequent references to atomic explosion and futuristic annihilation/salvation, the cartoons that dominate Japan's adult media and entertainment industries provide a screen that both exaggerates and diminishes real history, which they function to suppress' (2005, 247). The Japanese history Munroe refers is that of World War II, to do with the 'distorted dependence on the U.S.' and Japan's inability to resolve 'the trauma of nuclear war, the devastation of defeat, and the unmoored, apolitical state that has emerged since' (2005, 246 and 247). Mainstream Japanese manga/anime similarly draw upon the twisted perspectives generated by a resonating history of defeat, trauma, dependence, and an episodic prosperity characterised by political stagnation. A classic remote-controlled giant robot in the anime *Iron Man No. 28*, for instance, displays its destructive power when controlled by villains. However, the machine itself is regarded as neutral. It is humans who turn it into an evil machine; the robot itself is not evil. In the original plot, the mega-humanoid was developed by the Japanese military as a remote-controlled super-weapon to save the Japanese Empire at the end of World War II. It is said that the creator of *Iron Man No. 28,* Mitsuteru Yokoyama, got his inspiration for the work's title from the experience of witnessing American B-29 bombers over Tokyo's sky during the war (Murakami 2005, 142). Though it is intended as a manga/anime for children, *Iron Man No. 28* is not innocent of the lessons of Japan's ultimate defeat in wartime twinned with admiration and fear concerning American technological success. Astro Boy also embodies this twisted duality.

ASTRO BOY SYNDROME

The title of roboticist Hitoshi Matsubara's book can be translated as *Would we be able to actualise Astro Boy?* (1999). It is about Matsubara's work, centring on RoboCup, a robot competition event. (I will discuss RoboCup in Chap. 7). Despite the title, references to Astro Boy are made only briefly, at the beginning by Matsubara and at the end of the book in an afterword by writer Madoka Tainaka, who is one of the producers of the book and who interviewed Matsubara. Why reference Astro Boy? It is obvious that the provocative title was chosen for marketing reasons. The book aims to take advantage of Astro Boy's popularity and the robot's birthday in 2003. However, the book's title also suggests a deeper resonance. Whenever the Japanese public sees a robot, it is difficult not to refer to Astro Boy. The pervasive influence of Astro Boy is such that it is often used as a reference point for newly developed humanoids. Sena regards the close relationship between Astro Boy and the development of the humanoid in Japan as 'the curse of Astro Boy' (Sena 2008, 360). Sena means that the general public expects too much of humanoids as they currently are, seeing them in relation to Astro Boy, and the public becomes disappointed with the gap between reality and fantasy.

Astro Boy in its initial iteration was conceived as a mediator in several ways. Tezuka created the child robot character as a cartoon representation of the real need to navigate between two cultures, an idea based on his experiences with Americans in the post-war period: he was beaten by a drunken American soldier because of a language issue (Yonemura 2004, 78). Solving all such problems, Astro Boy is able to speak sixty languages. In Tezuka's view, Astro Boy is 'a kind of interface' between humans and machines (quoted in Sena 2001, 270). Tezuka states that through the story of Astro Boy, he wanted to highlight the lack of communication between ordinary humans and scientists (1996, 26). Yet when Astro Boy tries to learn to be a human, he is ostracised as an outsider, a non-human robot. Paradoxically, in the narrative, the boy robot's situation allows human beings to relearn what it is to be human (Inuhiko Yomota, quoted in Natsume 1995, 144). It is highly symbolic that *Astro Boy*'s original Japanese title in 1951 was *Atomu taishi* (Ambassador Atomu [atom]) before it was changed to *Tetsuwan* (Iron arm) *Atomu* in 1952. More importantly, Astro Boy is equipped with a nuclear-powered engine, reflecting the 'optimistic idea' regarding 'peaceful uses of nuclear energy', which was promoted in the 1950s (Tanaka 2010, 8).

Astro Boy's struggle for his own subjectivity resonates with Japan's own search for direction after the war. The view that Japanese anthropomorphic robots such as Astro Boy are something between the machine and the human suggests a mediating of Western modernity and technology in the post-war period. Japanese imaginary robots in manga/anime are 'an exuberant assertion of a collective technological fantasy' for the Japanese people (Schodt 1988, 73). Frederik Schodt summarises *Astro Boy*'s success and its impact on the Japanese popular psyche: 'Atom became the little boy next door, except he lived in the future where science and technology had created a world of clever gadgets and a standard of living that Japanese could only dream about' (Schodt 1988, 76). Since Atom is regarded as 'a child of science', according to Schodt, 'he – and robots – became linked with a wonderful future that science and technology could provide' for post-war Japan (Schodt 1988, 76). Astro Boy immediately became the symbol of the times, a period of great economic growth when Japan was rebuilding itself with new technologies in the decades after World War II. National milestones included the first television broadcast in 1953, the construction of Tokyo Tower in 1957, the first highway for automobiles in 1963, and the Shinkansen (bullet train) in 1964.

There is another respect in which Astro Boy is an important mediator and which also establishes the character's importance in the Japanese psyche. Sena feels that Japanese humanoid robotics explores robots as intermediaries between the organic and the inorganic, between life and the world of things (2008, 484). Masahiro Mori holds the view that if the robot is to approximate a human, the robot should be regarded as 'the third existence' (2014, 86). For Minoru Asada, robots should be regarded as 'robo-species' (Asada and Higaki 2013, 8), something sentient but between humans and machines. For Japanese humanoid researchers, Astro Boy was an early and prime mediating figure in this way, an empathetic, strong, and intelligent dream-machine (as in ideal machine, and as a cultural machine that generates fantasies and representations) that also replicates human functionality.[7]

Interestingly, filmmaker Mamoru Oshii sees Astro Boy as analogous to the World War II Japanese imperial navy battleship *Yamato*, which continues to capture the popular imagination for the Japanese to the present (quoted in Sena 2001, 263).[8] Importantly, Oshii sees both battleship *Yamato*, an actual physical entity that existed in Japanese history, and Astro Boy, a fictional character from manga/anime, in similar terms. What Oshii stresses is the uniqueness of these potent symbols, representing that which is not mass-produced but is a carefully crafted object reflecting the most advanced science and technology of the day. These unique and special machines are the opposite of the mass-manufactured B-29 bomber, which devastated Japanese cities in wave after wave of attack. Oshii discusses the concept of uniqueness in relation to the Japanese term '*yorishiro*', which describes objects associated with divine spirit. In other words, for Oshii, both battleship *Yamato* and Astro Boy possess a certain cultural resonance and force because they are technologies endowed with special spirit and uniqueness that seem to formulate a particularly Japanese address to technologized modernity. What is also important here, in the application of this concept of *yorishiro* to Astro Boy, is that attributes of cultural potency, iconicity, and, for the Japanese, spirit, infuse what I will discuss as the 'presence' of robots in robot performance.

THE PRESENCE OF THE JAPANESE HUMANOID

Manga/anime characters like Astro Boy have a particular significance for the Japanese, for the reasons I discussed above. There is a particular kind of iconicity that relates to characters within the worlds of manga and anime: it is '*kyara*', a word derived from 'character'. *Kyara* is in effect a highly condensed symbol.

In examining the history of Japanese manga, critic Eiji Otsuka argues that Tezuka is the first after World War II to introduced human characteristics of interiority to manga characters (2009, 178). Otsuka also points out the significance of Tezuka's introduction of mortal characters who perish in manga, unlike, for example, Disney characters who never die (2009, 137–41). Otsuka attributes this to Tezuka's experience of the American bombing of Osaka during the war. For Otsuka, these manga characters are paradoxical because they are to be understood simply as codes without human characteristics and hence should not express mental suffering or experience death as such. Manga critic Go Ito, on the other hand, distinguishes Japanese manga characters with interiority in the Tezuka style from the ones of *kyara* (as proto-characters) in pre-war times (Ito 2005, 111). For Ito, Tezuka's manga characters are regarded as human-like, with personalities, and their stories reflect life and living. A *kyara*, on the other hand, is much more simply drawn, and is often caricature-like. A *kyara* can be described with 'machine-like' characteristics such as being bound to rules, lacking interiority, and the absence of suffering (A. Kubo 2015, 75). Yet for serious manga fans, a popular *kyara* in its contemporary forms can offer a certain human-like presence in itself, without being imbedded within a manga story (Ito 2005, 263), and, according to Akinori Kubo, because of the tradition of animism, even a '*kyara*', a figure of a simple line drawing, can be sensed to have vitality or life force (2015, 73). Astro Boy reflects the dual structure of being a robot, a technological object, and being an icon (having the same power as *kyara*): doubly ambivalent, situated between the human and the non-human.

In the shadow of Astro Boy, the humanoid as both cutting-edge technology and fantasy is a loaded object and sign, with the capacity to facilitate positive feelings, expectation, or wonder for its Japanese viewers. For example, the appearance of ASIMO is based on the image of an astronaut in a spacesuit (Masato Hirose in Yonezawa 2002, 58), which is itself an icon for futurity and a benignly advanced technology. After reviewing the previous model P2 (182 cm high, 210 kg in weight), which was seen as menacing and threatening in close proximity, it was determined that ASIMO should have a child's height. ASIMO's head is helmet-like and its face is covered with a dark screen, while two large round eyes are recognisable behind the screen. Masato Hirose explains that he learnt from feedback that its ambiguous facial features, in contrast to the human face, would facilitate the imagination, which would contributes to the robot's popularity (cited in Yonezawa 2002, 59).[9] Less-realistic design approaches that see robots more in terms of *kyara* are often chosen in order to avoid falling into the 'uncanny valley'. It is possible to see in the abstract features of ASIMO an appeal to the auratic and iconic nature of *kyara*. For scriptwriter Yoshiki Sakurai, ASIMO's design emphasises its 'robot identity', distinguishing it from the human, and, at the same time, its child-like appearance eases people's expectations regarding ASIMO's mobility, as one can tolerate the untutored abilities of a child (2007, 133 and 137). ASIMO can be seen as a kind of robot *kyara* in its mix of attributes of familiar images of robot manga/anime characters.

Kyara are condensed symbols, and for fans of these characters can also radiate life force. ASIMO has been designed with these two aspects of *kyara* in mind. We might call the life force auratic. How the audience might relate to ASIMO and other humanoid robots can be explained in terms of theatre studies scholar Cormac Power's notion of 'the auratic mode of presence' rather than an actor's 'literal mode of presence' which is the immediate corporeal presence of the actor (2008, 87). Power discusses three modes of presence: the auratic, the fictional, and the literal. Power's 'fictional mode of presence' occurs when an object or person is made meaningful through a narrative storyline in which it is embedded (2008, 15). The auratic mode of presence, in its first iteration, is often understood in terms of 'the fame or reputation of the actor, playwright or artwork, along with the knowledge and expectations that spectators may carry with them into the experience' (Power 2008, 47).[10]

This notion alone is not appropriate because a humanoid, unlike an actor in a play, does not have interiority. It needs to be activated at a show, or in some context. In this sense, it is useful to consider Power's second notion of the auratic mode of presence, which arises not from fame or expectations but from the totality of effects created by the interconnections among actor, audience, and context: 'an idea of aura as created *through* the act of representation' (2008, 52, original emphasis). Aura in this sense is a result of a total representational strategy: 'the cultivated craft of the performers, the cultural context surrounding a theatrical performance, and the way in which a performance is itself represented to the audience' (Power 2008, 84). A connection between stage and audience is facilitated through the audience's conscious involvement. In other words, this sense of aura can be felt when the audience experiences the performance through 'a heightened state of awareness' of the meaning of the total performance that occurs somewhere between stage and audience (Power 2008, 82). Indeed, ASIMO became famous not as a static object but as a performing object in staged demonstrations. In general, the mimetic humanoids are enabled for the audience through carefully staged enactments. While humanoids can be felt as *kyara* to fans, as is ASIMO, the auratic quality of the humanoid is necessarily the result of its staging. ASIMO's demonstrations are designed in a way that encourages the viewers to appreciate the auratic presence of the desired robot while masking its technical limitations.

In the following chapters, I examine specific instances of complex theatrical events that feature Japanese mimetic robots. The stage directions and dramaturgies of these events were designed to set up appropriate *mises en scène* that facilitate the 'auratic' mode of presence of the mimetic robots. So often in these demonstrations and shows, the politics of the humanoid and android is played out: the humanoid is staged in ways that, as I will argue, further positive themes to do with nationalism, corporate identity, and the broader theme that I have explored in this chapter to do with robots' role in envisioning a positive and prosperous technologized future.

An understanding of the auratic notion of presence helps to highlight the aspects of robot performance that are highly dependent upon context. I will apply these different understandings of presence to examine robot performances in the next four chapters. I will also return to the strange affective powers of the humanoid in relation to its representational contexts in later chapters, as in my discussion of Oriza Hirata's plays and Kenji Yanobe's installations. The next chapter examines large-scale contemporary robot events, and those of the 1970s, where humanoid robots are spectacularly displayed.

NOTES

1. Later models are equipped with sound and voice recognition systems and speak to humans. They can shake hand with a human interactant and carry a food tray. The 2011 model of ASIMO is able to walk much faster, and even run (9 km/h). It can climb up and down the stairs as well as jump and hop on one leg. When moving, it can avoid collisions with humans, predicting their movements. This model is also able to open a PET bottle and pour into a paper cup.
2. The other participating organisations involved in developing the humanoid include the government agency National Institute of Advanced Industrial Science and Technology (AIST), Yaskawa Electric, and Shimizu Corporation. In the application phase of the HRP project, other major companies—such as Honda, Kawasaki Heavy Industries, Fanuc, Fujitsu, Matsushita Electric Industrial (current Panasonic Corporation), Hitachi, and Mitsubishi Heavy Industries—also participated.
3. Honda's English-language webpage describes the ASIMO project, not in terms of its vision for the future but in terms of the company's pursuit of new technologies for mobility and their applications (Honda Motor Company 2015b).

4. In Japan, it is common to see journalistic and semi-academic publications on topics in areas typically treated by academic disciplines.

5. There were reported accidents, including deaths of workers, at several nuclear power plants in Japan before the Fukushima disaster: at the Mihama nuclear power plant (1991 and 2004), at the Monju prototype fast-breeder reactor (1995 and 2010), and at the Tokaimura nuclear fuel-processing plant (1997 and 1999). The need for robots equipped to navigate hazardous environments was acknowledged (Sena 2004, 598).

6. Asimov's three laws are as follows: 'One, a robot may not injure a human being, or, through inaction, allow a human being to come to harm.... Two, ...a robot must obey the orders given it by human beings except where such orders would conflict with the First Law.... And three, a robot must protect its own existence as long as such protection does not conflict with the First or Second Laws' (Asimov 2004, 37, my ellipses).

7. Japanese humanoid researchers often refer to their research methodologies as 'constructivist approaches', that is, approaches structured by 'learning by doing' (Ishiguro 2007, 116; Inamura and Ikeya 2009, 63; Kuniyoshi 2008, 72; Taniguchi 2010, 29; Asada 2010, 25). Just as humans learn by doing, the researchers create robots that can learn through trial-and-error. Through that process, roboticist Yasuo Kuniyoshi, for example, aims to model the entire process of the physical and psychological development of the human through robots (2008, 73–9). For Ishiguro, a group of robots is used to simulate social communication through non-verbal means in order to model human sociality (Ishiguro and Washida 2011, 141).

8. Battleship *Yamato*, along its sister ship *Musashi*, was armed with the world's heaviest and most powerful weapons and yet was sunk by US torpedo bombers, on its way to a suicide mission in Okinawa, without fulfilling its mission. A popular science fiction anime, *Space Battleship Yamato* (1974), which, in the story, is built from the fragments of the actual battleship, continued to run in a television series and in later films and related products.

9. Discussions on the look of ASIMO reveal Honda's concerns for its mainstream appeal. In contrast, US firm Boston Dynamics—a leading robotics company known for the development of quadruped robot BigDog, and more recently, Atlas, a 1.8-metre bipedal humanoid, both of which are capable of walking in an open field—is focusing

on robots' functionality and is less concerned with the 'friendly' appearance of robots as such.

10. For other studies on the actor's auratic presence, see Roach (2007) and Goodall (2008).

References

Asada, M. (2010). *Robotto to iu sisō* [A philosophy called the robot]. Tokyo: NHK Shuppan.

Asada, M., & Higaki, T. (2013). Robotto, ningen, seimei [Robot, human being, life]. In T. Higaki (Ed.), *Robotto, sintai, tekunolojī : baiosaiensu no jidai ni okeru ningen no mirai* [The robot, the body, technology: The future of the human being in the age of bioscience] (pp. 3–35). Osaka: Osaka Daigaku Shuppankai.

Asimov, I. (2004). *I, Robot* (Bantam hardcoverth ed.). New York: Bantam Dell.

Baba, N. (2004). Honsho no nerai [The objective of this book]. In *Robotto no bunkashi: kikai o meguru sō zō ryoku* [Cultural analysis on the robot: The imagination concerning machines] (pp. 7–11). Tokyo: Shinwasha.

Breazeal, C. L. (2002). *Designing sociable robots*. Cambridge, MA: MIT Press.

Doi, T. T. (2012). Inugata robotto AIBO to shin robotto sangyō [Robot Dog AIBO and the New Robot Industry]. *Nihon Robotto Gakkaishi, 30*(10), 1000–1001. doi:10.7210/jrsj.30.1000.

Furuta, T. (2006). 2sai no korokara robotto hakase: Robotto gijutsu de bunmei, bunka no shinpo ni kōken sitai [Dr Robot from the age of two: Wanting to contribute to the progress of civilisation and culture through robot technology]. In P. H. P. Kenkyūjo (Ed.), *Otonano tameno robotto gaku* [Study of the robot for adults] (pp. 11–38). Tokyo: PHP Kenkyūjo.

Geraci, R. M. (2010). *Apocalyptic AI visions of heaven in robotics, artificial intelligence, and virtual reality*. Oxford/New York: Oxford University Press.

Goodall, J. R. (2008). *Stage presence*. London/New York: Routledge.

Hirose, M. (2006). Nirin, yonrin, soshite kyūkyoku no idōtai e [Two wheels, four wheels, and the ultimate vehicle]. In P. H. P. Kenkyūjo (Ed.), *Otonano tameno robottogaku* [Study of the robot for adults] (pp. 113–141). Tokyo: PHP Kenkyūjo.

Hirose, S. (2011). *Robotto sō zō gaku nyūmon* [Introduction to a study on robot creation]. Tokyo: Iwanami shoten.

Hirose, M., & Ogawa, K. (2007). Honda humanoid robots development. *Philosophical Transactions of the Royal Society of London A: Mathematical, Physical and Engineering Sciences, 365*(1850), 11–19. doi:10.1098/rsta.2006.1917.

Honda, Y. (2014). *Robotto kakumei: naze gūguru to amazon ga tō shi suru noka* [The robot revolution: Why Google and Amazon have invested in it]. Tokyo: Shōdensha.

Honda Motor Company. (2015a). Honda ASIMO robotto kaihatsu no rekishi [The history of ASIMO's development]. http://www.honda.co.jp/ASIMO/history/. Accessed on 10-10-2015.

Honda Motor Company. (2015b). About Honda Robotics. http://world.honda.com/HondaRobotics/. Accessed on 10-10-2015.

Hornyak, T. N. (2006). *Loving the machine: The art and science of Japanese robots* (1st ed.). Tokyo/New York: Kodansha International.

Hotta, J. (2008). *Hito to robotto no himitu* [The secret concerning robots and humans]. Tokyo: Kōbunsha.

Inamura, T., & Ikeya, R. (2009). Nōkagaku to robotikusu [Brain science and robotics]. In T. Inamura, H. Sena, & R. Ikeya (Eds.), *Robotto no oheso* [The robot's navel] (pp. 48–65). Tokyo: Maruzen.

Ishiguro, H. (2007). *Andoroido saiensu: Ningen wo sirutame no robotto kenkyū* [Android science: A study to learn what the human is]. Tokyo: Mainichi Komyunikēshonzu.

Ishiguro, H. (2009). *Robottoto wa nanika: Hito no kokoro wo utsusu kagami* [What is the robot?: A mirror reflecting the human soul]. Tokyo: Kōdansha Gendaishinsho.

Ishiguro, H., & Washida, K. (2011). *Ikirutte nanyaroka: Kagakusha to tetsugakusha ga kataru, wakamono no tame no kuritikaru jinsei sinkingu* [What does it mean to live: A conversation between a scientist and philosopher, critical life thinking for young people]. Tokyo: Mainichi Shimbunsha.

Ishihara, N. (2011, April 28). Genpatsu, anzen shinwa ni manshin sita tsumi [Nuclear power plants, A crime caused by a myth of safety]. *Nikkei Business Online*. http://business.nikkeibp.co.jp/article/manage/20110426/219655/. Accessed on 05-15-2012.

Ito, G. (2005). *Tezuka izu deddo: Hirakareta manga hyōgenron e* [Tezuka is dead: Open expression in manga]. Tokyo: NTT Shuppan.

Kajita, S. (2008). Hyūmanoido robotto kenyū no genba yori [On the ground with humanoid research]. In S. Hideaki (Ed.), *Saiensu imajinē shon: Kagaku to SF no saizensen, soshite miraie* [Scientific imagination: The frontline of science and SF, and the future] (pp. 38–53). Tokyo: NTT Shuppansha.

Kakoudaki, D. (2014). *Anatomy of a robot: Literature, cinema, and the cultural work of artificial people*. New Brunswick/New Jersey/London: Rutgers University Press.

Kobayashi, H. (2006). *Robotto sinkaron: Jinzō ningen kara hito to kyōzon suru sisutemu e* [Robot evolution: From Jinzōningen to a system for coexistence with humans]. Tokyo: Ōmusha.

Kubo, A. (2015). *Robotto no jinruigaku: Nijūseiki no kikai to ningen* [Anthropology of the robot: The machine and the human in the 20th century]. Tokyo: Sekai Sisōsha.

Kuniyoshi, Y. (2008). Robotto bodei robotto maindo: Mono no sekai to kokoro no sekai no yūgō soshite kaihatsu [Robot body, robot mind: Developing the integration of the worlds of soul and object]. In H. Sena (Ed.), *Saiensu imajinē shon: kagaku to SF no saizensen, soshite miraie* [Scientific imagination: The frontline of science and SF, and the future] (pp. 70–87). Tokyo: NTT Shuppansha.

Matsubara, H. (1999). *Tetsuwan Atomu wa jitsugen dekiruka: Robo kappu ga hiraku mirai* [Would we be able to actualise astro boy?: The future created by RoboCup]. Tokyo: Kwade Shobō.

Matsubara, H. (2001). Robo kappu no yume [A dream of RoboCup]. In H. Matsubara, I. Takeuchi & H. Numata (Eds.), *Robotto no jō hō gaku: 2050 nen wā rudo kappu ni katsu* [Robot information study: Winning the World Cup in 2050] (pp. 7-66). Tokyo: NTT Shuppansha.

Mori, M. (2014). *Robotto kō gaku to ningen: Mirai no tameno robotto kō gaku* [Robotics and the human: Robotics for the future]. Tokyo: Ōmusha.

Munroe, A. (2005). Introducing little boy. In T. Murakmi (Ed.), *Little boy: The arts of Japan's exploding subculture* (pp. 241–261). New York/New Haven/London: Japan Society and Yale University Press.

Murakami, T. (2005). Earth in my window. In T. Murakami (Ed.), *Little boy: The arts of Japan's exploding subculture* (pp. 99–149). New York/New Haven/London: Japan Society and Yale University Press.

Nagata, T. (2005). *Robotto wa ningen ni nareruka* [Can a robot be a human being]. Tokyo: PHP Kenkūsho.

Natsume, F. (1995). *Tezuka Osamu wa dokoni iru* [Where is Osamu Tezuka]. Tokyo: Chikuma Shobō.

Nishida, T. (2005). Riaru ējento to siteno robotto [A robot as a real agent]. In H. Inoue, T. Kanede, M. Uchiyama, M. Asada, & Y. Anzai (Eds.), *Iwanami kō za robottogaku 5: Robotto infomatikkusu* [Iwanami robot science 5: Robot informatics] (pp. 89–139). Tokyo: Iwanami Shoten.

Orbaugh, S. (2005). The genealogy of the cyborg in Japanese popular culture. In K.-y. Wong, G. Westfahl, & A. K.-s. Chan (Eds.), *World weavers globalization, science fiction, and the cybernetic revolution* (pp. 55–71). Hong Kong: Hong Kong University Press.

Otsuka, E. (2009). *Atomu no meidai: Tezuka Osamu to sengo manga no shudai* [An atom thesis: Osamu Tezuka and themes in Postwar Manga]. Tokyo: Kadokawa shoten.

Power, C. (2008). *Presence in play a critique of theories of presence in the theatre.* Amsterdam/New York: Rodopi.

Roach, J. R. (2007). *It.* Ann Arbor: University of Michigan Press.

Sakurai, Y. (2007). *Firosofia robotika: Ningenni chikazuku robotto ni chikazuku ningen* [Philosophia robotica: Humans who become similar to robots that approximate humans]. Tokyo: Mainichi Komyunikēshonzu.

Sankai, Y. (2006). Shintaikinō wo kakuchō suru: Robotto sūtsu de hito no yakunitatsu [Expanding physical capabilities: A robot-suit that is useful for humans]. In P. H. P. Kenkyūjo (Ed.), *Otonano tameno robotto gaku* [Study of the robot for adults] (pp. 93–114). Tokyo: PHP Kenkyūjo.

Schodt, F. L. (1988). *Inside the robot kingdom: Japan, mechatronics, and the coming robotopia.* Tokyo/New York: Kodansha International.

Sena, H. (2001). *Robotto Seiki.* Tokyo: Bungei Shunjū.

Sena, H. (2004). In H. Sena (Ed.), *Robotto Opera (Robot Opera: An anthology of robot fiction and robot culture, original English title).* Tokyo: Kōbunsha.

Sena, H. (2008). *Sena Hideaki robottogaku ronshū* [Hideaki Sena robot study essay collection]. Tokyo: Keisō Shobō.

Sugano, S. (2011). *Hito ga mita yume, robotto no kita michi: Girisha shinwa kara Atomu, soshite* [Mankind's Dream, The Historical Path: The Robot From (ancient) Greece to Astro Boy]. Tokyo: JIPM Sorūshon.

Takahashi, N. (2011). *Bō karoido genshō : Sinsē ki kontentsu sangyō no mirai moderu* [The phenomenon of Vocaloid: The future model for content business in the new century]. Tokyo: PHP Kenkyūjo.

Takeno, J. (2011). *Kokoro wo motsu robotto: Hagane no shikō ga kagami no nakano jibunni kizuku* [The robot with soul: A realisation of the self in the mirror by a metallic thinking entity]. Tokyo: Nikkan Kōgyōsha.

Tanaka, Y. (2010). War and Peace in the Art of Tezuka Osamu: The humanism of his epic manga. *The Asia-Pacific Journal: Japan Focus, 8*(38.1), 1–15.

Taniguchi, T. (2010). *Komunikeishon suru robotto wa tsukureruka* [Is it possible to build a communication robot]. Tokyo: NTT shuppan.

Taylor, C. (2004). *Modern social imaginaries.* Durham/London: Duke University Press.

Tezuka, O. (1996). *Garasuno chikyū wo sukue: Nijū isseiki no kimitachi e* [Save the glass earth: For you, the 21st-century generation]. Tokyo: Kōbunsha.

The Humanoid Research Group of National Institute of Advanced Industrial Science and Technology. (2015). Hataraku ningengata robotto HRP-2 purototaipu wo kaihatsu [Developing working humanoid HRP-2, prototype]. https://unit. aist.go.jp/is/humanoid/m_projects/hrp-2p_j.html. Accessed on 09-15-2015.

Toyota Motor Corporation. (2015). Toyota Global Site, Expo 2005 Aichi. http:// www.toyota-global.com/innovation/partner_robot/aichi_expo_2005/index04. html. Accessed on 10-10-2015.

Yamada, N. (2013). *Robotto to nihon: Kingendai bungaku, sengo manga niokeru jinkō shintai no hyō sō bunseki* [The robot and the human: An analysis of representations of the artificial body in modern literature and Postwar Manga]. Tokyo: Rikkyō Daigaku Shuppan.

Yamato, N. (2006). *Robotto to kurasu: Kateiyō robotto saizensen* [Living with a robot: The frontline of domestic robots]. Tokyo: Sofuto Banku Kurieitibu.

Yonemura, M. (2004). Atomu Ideorogī [Atom ideology]. In N. Baba (Ed.), *Robotto no bunkashi: Kikai wo meguru* sōzōryoku [Cultural analysis on the robot: The imagination concerning machines] (pp. 74–105). Tokyo: Shinwasha.

Yonezawa, Y. (Ed.). (2002). *Robotto manga wa jitsugen suruka* [Could Robot Manga be Actualised]. Tokyo: Jitsugyō no Nihonsha.

CHAPTER 3

Futuristic Spectacle: Robot Performances at Expos

It is highly symbolic that Japan's first humanoid robot, Gakutensoku, performed at a national exposition held in the midst of Japan's transformation to a modern society in the early twentieth century. Gakutensoku set a precedent: a robot could be exhibited for the public as a captivating object of fascination, spectacularly showcasing Japanese science and technology. Indeed, the close relationship between the robot and the spectacle in large-scale, government-led exhibitions continues to this day in Japan. This chapter examines how robots were theatrically presented to enact patriotism at the international expositions that were held in post-World War II Japan. The expositions I consider include the Japan World Exhibition in Osaka in 1970, the International Exposition in Tsukuba in 1985, the International Garden and Greenery Exposition in Osaka in 1990, and the World Exposition in Aichi in 2005.

The robot has been one of the key themes of these expositions, manifesting the dreams of their Japanese audiences concerning their country's future. As I have discussed, the robot has signified technological advancement and prosperity for post-war Japan. The emphasis on the robot as a central figure of scientific and technological advancement in Japan, through successive generations, indicates what anthropologist Alfred Gell calls a 'technology of enchantment' (2009). Gell indicates that the social and symbolic meanings of technology become crystallised in an object, which then seems to embody the enchanting properties of that technology. Antiques may have no practical use but serve the purpose of signifying history, Jean Baudrillard suggests (2009, 41), and it is similarly the case

© The Author(s) 2017
Y. Sone, *Japanese Robot Culture*,
DOI 10.1057/978-1-137-52527-7_3

with advanced humanoids in Japan and their symbolic crystallisation of ideas to do with technology, futurity, and the related affects of hope and optimism. The robot performances at these expositions constitute a transformative theatrical space—and as such, there is room for affective response related to nationalistic sentiment to be conjured.[1]

Studies of the theatrical object reveal complex material operations in performance. The reconsideration of the subject-object relation in terms of theatrical space and objects has been an important area of inquiry in performance theory. For theatre studies scholar Gay McAuley, when the transactions between actors, spectators, objects, and the space itself are established in a given space in theatre productions, 'that space itself is activated and itself made meaningful' (1999, 8). Focusing on the materiality and sensory experience of the object in post-war theatre, Stanton B. Garner highlights instances when 'the object is freed from subordination in a system of instrumentality and looms increasingly as the thing itself' (1994, 109). Garner discusses the phenomenological effects of objects on stage in the works of prominent post-war playwrights such as Sam Shepard, Eugene Ionesco, Samuel Beckett, and Harold Pinter. In an issue of the theatre and performance studies journal *Performance Research* in 2007 that focused on such concerns, the authors question and problematise the subject-object relations of traditional plays by pointing out the elevated status of theatrical objects in contemporary Western theatre and performance works (Clarke et al. 2007).[2] Objects are more than simply props: they assume ritualistic functions, or they become invested with mnemonic power.

However, theatre studies' understanding of the object within theatre is not sufficient to address the performances of technological objects outside the practises of Western artistic theatre, for example, at international expositions in the Japanese context. First, it is important to understand that the expos themselves essentially present through the form of the 'exhibition' as well as through various 'performance genres: ritual, ceremony, drama, theatre, festival, carnival, celebration, [and] spectacle' (Roche 2000, 9). Each expo's theme also contributes a superstructure of meaning, as all its events are intended to be illustrative in some manner of the guiding theme. And, as I have already indicated, an analysis of robots as theatrical objects at Japanese expositions requires an awareness of their specific socio-political contexts, essentially their ideological content, and how this content facilitates transformative effects for visitors. I will articulate this latter point below, referring to what Shunya Yoshimi calls '*banpaku gensō*' (fantasies through the exposition) (2005). Yoshimi's discussion is useful in identifying the patriotic thrust of the Japanese expo, which was evident from its origins.

A survey of historical expositions in the West quickly reveals that whether in the West or in Japan, fantasies of the social are part of the attraction and appeal of expositions of this kind and that all expositions tend to emphasise themes that reflect the nation's history through culture as much as through science. For example, while industrial fairs became common in eighteenth-century Europe, the Great Exhibition of the Works of Industry of all Nations that took place in 1851 at the Crystal Palace in London was the first of many large-scale, spectacular international exhibitions in the nineteenth century in Europe (and in the USA), where self-images of powerful imperial countries were projected through the panoramic displays of manufactured products, industrial technologies, and exotic goods. Early twentieth-century expositions in the West included notorious 'human exhibits' from the colonies as ethnographic displays.[3] Later in the twentieth century, commercial advertisers and industrial designers began choreographing and staging spectacles of affluence and consumerism for massively scaled corporate pavilions that were themselves both monuments to American capitalist powers and advertisements for them (Yoshimi 1992, 252).

There is a correlation between spectacle and references to the future in the history of Japanese expositions. It is important to note that discussion of the future was already present at the birth of the modern government-led exhibition in Japan, as early as the late nineteenth century, following European examples. Japan's first National Industrial Exhibition in Tokyo was held in 1877 during the early stages of the country's modernisation and nation-building through rapid industrialisation and Westernisation. Modern Western technologies, and aspects of Western civilisation as a whole, were seen as models to follow to secure the future of the newly established Meiji Japan. The exposition was the new government's effort to establish a modern-style national exhibition, showcasing Japanese products in a modern way, as it was learned from the Vienna International Exposition in 1873 (National Diet Library 2010).[4] The exhibition was designed to bring a new, Western approach to exhibition display to Japan, and, importantly, it was also intended to bring with it the idea of objective viewing: a scientific viewing through distanced eyes. However, the idea of the objective, scientific gaze was foreign to the Japanese audience of that time, which was accustomed to entertainments from the pre-modern period. This kind of modern viewing, purporting to be objectively distanced from its subject, requires a separation of subject and object. On the other hand, the Japanese pre-modern viewing of spectacles, '*misemono*'—whether of Buddhist religious objects or of Kabuki performance—merged the viewers with 'exhibited' objects and players in a non-everyday space that was understood as linked phenomenologically to the divine world via the sense of 'touch', rather than privileging vision, and through a logic of 'infection' (Yoshimi 2008, 193). Pre-modern Japanese people flocked to gain divine favour at a popular showing of a Buddhist temple's or a shrine's treasures, artefacts, and curiosities (Yoshimi 2008, 147).

Later exhibitions began accommodating popular *misemono* spectacles, including stunt artists, acrobatics, sword performers, magic shows, or *karakuri* performances at or beside marketplaces or streets of shops. For example, an exhibition in 1903 included a carousel, a Ferris wheel, a water-chute ride, a panorama museum, a diorama museum, as well as a wonder museum that featured a blonde female dancer with colourful lights. The 1914 Tokyo Taisho Exposition was also very popular. It featured attractions such as a cable-car specially built for the exposition; a mummy of an Indian monk; a mining pavilion shaped like a large rock; the purportedly man-eating natives of the South Sea Pavilion; and the Beauty Island Pavilion, with its female performers costumed in the figures of snake, horse, or fire (Yoshimi 2008, 148–9). The theme-park-styled, curiosity-driven spectacle of this tradition of large-scale expositions in Japan, which persists to the present day, has its roots in the spirit of *matsuri*, festivals or carnivals dating back before pre-modern Japan. There were similar national exhibitions in Japan until the outbreak of World War II.

The first international expo in Asia after the war was held in Osaka in 1970. According to Yoshimi, the Japanese expos of the 1970s were the products of 1960s government programmes bent on 'doubling national income' and 'regional development' in the context of post-war politics (2005, 33). The Osaka expo set the precedent for later international exhibitions. Successive Japanese governments invested in the regional development of the exhibition sites, providing celebratory occasions for the Japanese people to share in the wealth of the nation. Importantly, robots performed in this first international expo in Japan.

Robot demonstrations at the international expositions held in Japan that I will discuss afforded excellent opportunities to promote Japan's technological achievements to both domestic and international audiences. The spectacular presentations of these expositions were crafted to make the robot enchant Japanese spectators. This chapter highlights how these robot spectacles furthered official narratives. Seeing such events in terms of their history as *misemono*-like popular entertainments allows us to understand the avid interest of the expo's visitors and indicates that their context, not only that of triumphal festival and joyous celebration but as deployments for other purposes, must be taken into account.

At the end of the chapter, I will also examine the Great Robot Exhibition, held in 2007 in Tokyo, as an excellent example of the rhetoric of *nihonjinron*, that is, Japanese essentialism, for the promotion of the next-generation robot. I examine the dramaturgy of ASIMO's performance skit in the Great Robot Exhibition. In the Great Robot Exhibition, ASIMO was essentially a character in a story, while, when promoted in Sydney, ASIMO danced in a much simpler demonstration spectacle for the Western audience. I consider why, with regard to *nihonjinron*, ASIMO appears so differently to these two audiences.

Robot Spectacle at the Japan World Expo, 1970

Japan's post-war recovery led to a period of high economic growth from the mid-1950s. As discussed in the previous chapter, Astro Boy, created in 1951, became the symbol of the period, and its popularity led to an anime series in 1963. This was also the time that American industrial robots were first introduced to Japan. Unimate was the world's first industrial robot, developed by American inventor George Devol and physicist and engineer Joseph Engelberger in the late 1950s. Japanese engineers frequently visited Devol and Engelberger's company Unimation in the USA in the early 1960s. Kawasaki Heavy Industries established a licencing agreement with Unimation to produce their industrial robots, to be used for spot-welding in car manufacturing factories, and began producing industrial robots in the late 1960s. Japan's export boom of the period, led by the automobile and electronic appliance industries, necessitated automation of manufacturing on a massive scale. Automation and robotisation were seen as essential to the Japanese wealth-generation formula: importing raw materials and exporting industrial goods.

The Japan World Exposition in Osaka in 1970 represented Japan's post-war recovery and growing economy in terms of the promise of a secure economic future and endless prosperity. According to art critic Noi Sawaragi, the 1970 World Exposition channelled the nation's energy for the longer recovery period, directed toward an 'optimistic futurism and scientism' (2005, 115–6). As Sawaragi observes, the expo was essentially an apparatus for a nationalistic ideology, in which the progress of humanity was guaranteed through technological innovations (2005, 44), and the promise that Japan would be well placed to offer the world those innovations going forward.

The exposition was also an occasion to display evidence of the country's successful and productive modernisation since the late nineteenth century (Sawaragi 2005, 148). Scholars remark that the 1970 exposition was a realisation of the postponed plan for an international exposition during the war, which was supposed to be part of the celebration to mark the year 2600 after the enthronement of Emperor Jinmu in 660 BC (Yoshimi 2005, 43; Sawaragi 2005, 148). Public money poured into the 1970 expo, as it was planned by the political and business establishments as an important national event after the Tokyo Olympics in 1964 to represent the country's post-war successes. For Sandra Wilson, a historian of modern Japan, '[t]he Tokyo Olympics of 1964 and Expo '70 in Osaka were the most self-conscious displays of "nation" in Japan since the end of the Second World War' (2012, 159). Accordingly, it was set to be an event of nationhood, as its formal title used 'Japan' instead of the name of the city, 'Osaka', which was the first time this approach to naming occurred in the history of the World Exposition (Yoshimi 2005, 40). Because of the national importance of the event, it was essential to locate it in relation to both the business world and Japan's intelligentsia. Many prominent academics and artists were involved in the planning and production stages of the exposition (Yoshimi 1992, 223).[5] All outlets of the Japanese mass media encouraged the Japanese people to attend the event. Powerful non-government organisations, such as the Japan Agricultural Cooperatives and Japanese National Railways, mobilised their members to travel to the exhibition (Yoshimi 1992, 229). Records indicate that 64 million visitors attended, which was the highest number until the Shanghai World Expo, which had 73 million people visiting it in 2010.

Related to the main theme of progress and harmony for mankind, *omatsuri* (honorific '*o*' and *matsuri*) was chosen as the key vision for the exposition. Architect Kenzo Tange, one of the master planners, conceived of the exposition as a festival, gathering people, communicating people's energies, and as a vehicle for the exchange of intelligence and creativity between people (Yoshimi 1992, 224). Artist Taro Okamoto, one of the event's key producers and the creator of The Tower of the Sun, a building known as the symbol of the exposition, saw the essence of the exposition in terms of a feeling of exaltation with surprise and rejoicing, the antithesis to modalities of scientific knowledge, as such (Yoshimi 1992, 224). Pavilions and their displays were filled with the latest technologies of light and sound, especially the projection technologies of slide and film, in order to celebrate, in literal terms, a bright future. Performing robots played a prominent role. These machines included robot inventor and promoter Jiro Aizawa's entertainment robots at Fujipan robot pavilion and architect Arata Isozaki's gigantic box-shaped robots for the Festival Plaza, the main public entertainment space of the exposition.

Jiro Aizawa was known as 'Dr Robot', an engineer who began building entertainment robots since 1925 (Schodt 1988, 198) and published numerous robot engineering books in the pre-war years (Inoue 2007, 76). Through the Japan Institute of Juvenile Culture, which he established in 1952 with support from Osamu Tezuka as well as Masaru Ibuka, the cofounder of Sony, Aizawa produced many entertainment humanoids, around 800 in total (Moriyama 2009). Aizawa's robots all look like a tin toy robot with a square head and body, and two large box-shaped feet. These humanoids were of human height, though the tallest one was more than two metres high. They were remotely controlled and able to move arms, heads, and legs and equipped with a recorded voice. Aizawa regarded the humanoid as a human being, giving each one Japanese boys' names, such as Ichiro or Saburo. He regarded them as 'brothers'. The aim of the institute was to contribute to the welfare of children through the use of science-based toys. These robots were dispatched to museums and department stores all over Japan for entertainment purposes. At the 1970 Osaka exposition, Aizawa's robots were exhibited at the Fujipan robot pavilion, which Tezuka produced. Aizawa created many colourful robots that simulated dancing, singing, and playing musical instruments. There was even a robot taking photos of the visitors posed with another robot. The staging inside the pavilion was fantastical, representing a child's vision of entering the world of tin robot toys. Aizawa's robots were very well received and very popular for both children and adults (Moriyama 2009).

Arata Isozaki's robots, in contrast, were two gigantic 14-metre-high machine monsters that had rectangular bodies of approximately five metres in width and five metres in depth. One of them was called 'Deme', meaning 'pop-eyes' in Japanese and the other, 'Deku', a term coined by Isozaki from '*dekunobō*', a derogatory term for 'an incompetent and stupid person' (Isozaki, interviewed by Sawaragi 2002, 11). Deme was the main performer, and Deku functioned as the power supply. For Isozaki, the names Deme and Deku were meant to be friendly nicknames to anthropomorphise the giant machines that were not that sophisticated due to the technological limitations of the day (interviewed by Sawaragi 2002, 10). Deme has two arms, with the head section a T-shaped beam and a cube-shaped cockpit to house human operators at each end. Deme was equipped with sensors to collect atmospheric data and, following the operator's instructions, it illuminated its lights, and emitted sound and smoke. It was able to raise its body by its four legs attached to its bottom part.

Deme and Deku were positioned at the Festival Plaza square under the enormous roof (30 metres high, 290 metres long, 108 metres wide) designed by Tange, a grand and imposing setting. The plaza also housed the 70-metre-high Tower of the Sun, a huge, anthropomorphic structure protruding from the roof: quite the situation to impress viewers with its scale and a form that was at the same time both person and structure, not unlike its large-scale robots. As an anthropomorphic figure, the large scale of the Tower of the Sun echoed that of the very large, imposing, blocky shapes of Deme and Deku. The Tower of the Sun had three faces, representing present, past, and future. The face of the present was white, while the face of the future was gold, signalling a rich and promising future, and that of the past was black, signalling the metaphoric darkness of the recent past (Osakafu Nihon Banpaku Kinen Kōen Jimusho [Expo '70 Commemorative Park Office] 2014). The immense Tower itself was gleaming white, and the face of the present was surrounded by red rays, like the rays of the sun. Japanese visitors, looking up into the roundel of the face of the present, would have been reminded of the hope and renewal that the sun signifies in Japanese culture, and, of course, of the Japanese flag depicting the rising sun.

Despite his grand plans, Isozaki was disappointed with the dramaturgical dullness of his robots' performance at the opening ceremony of the exposition, held at the Plaza, which the Emperor Hirohito attended. After the opening show, Deme and Deku were not used as much as he hoped for, despite Isozaki's efforts, working for five years on an innovative architectural design for the space in which the machines would be displayed, in collaboration with the young artists of his generation.[6]

These two robot events were indicative of the nature of Expo '70 in Osaka. The expo was politically motivated, highlighting the success of postwar Japan and its corporate power. It was delivered and perceived as a large-scale, popular spectacle. Aizawa's robots and Isozaki's Deme and Deku were at the same time friendly performers and clowns and imposing spectacles placed in prominent locations. Even the emperor visited the Festival

Plaza where Deme and Deku were located. They were hilarious or grand or fascinating distractions, replacing any thought visitors may have had of the controversial political and social issues of the 1960s, such as the student activism of the era and the memory of the war, with *matsuri*-like sentiments of enjoyment. Visitors were awed by the sheer physical scale and conspicuously abstracted forms of its buildings and plazas, emblems of 'the modern' in the West. They appreciated Japan's achievements, and shared a common dream for the future. Theatre and performance studies scholar Peter Eckersall succinctly summarises the expo as 'a panacea to transform social anxiety and unrest into a positive futurist vision' of cleanliness and efficiency (2013, 121). Its robot events set the tone for robot performance events at national and international exhibitions in Japan. The popular appeal and style of Aizawa's performing robots become the standard for robot appearances at expositions, while Isozaki's experimental approach survives in a fine arts context. (I will discuss Kenji Yanobe's robot exhibits in the next chapter in relation to Isozaki's robots.) The 'acceptable' image of the robot as a sign of a bright future became the norm for robot performance in successive corporate-led international exhibitions in Japan.

CELEBRATING JAPAN'S ROBOTS: THE TSUKUBA EXPOSITION, 1985

The 1970s saw the rapid growth of the robot industry, with major corporations such as Fujitsu Fanuc, Fuji Electric, and Yaskawa Electric starting to develop their own industrial robots. The Japan Industrial Robot association (JIRA), the world's first robot industry association, was formed in 1971 (Engelberger 1999, 7). In the late 1970s, Kobe Steel, with Toshiba, developed a new type of assembly robot. The year 1980 is regarded as a milestone, as car production numbers in Japan exceeded those of the USA, due to the robotisation of Japanese car manufacturing plants (Sena 2001, 161). The Ministry of International Trade and Industry (now the Ministry of Economy, Trade and Industry) declared 1980 as the first year that Japanese industrial robots were widely used (Sugano 2011, 106). In 1983, the Robotics Society of Japan, a peak body for Japanese roboticists, was formed. It is not surprising that the theme of the 1985 Tsukuba Exposition, 'Humans, Dwellings, Environment, Science and Technology', would reflect the rapid development of the Japanese robot industry. Expo '85 was described as a 'high-tech Disneyland' (Haberman 1985). It was perceived as a spectacle to promote the government's policy on science and technology, and a variety of performing robots, not only industrial robots, featured at the expo, targeting children. Robots were presented alongside computers and new visual and projection technologies (Yoshimi 2005, 172).

Reflecting the growth of the robot industry in Japan, industrial robots were used at the pavilions of the 1985 exposition for performance purposes. For example, Fujitsu Fanuc displayed their giant 'Fanuc Man', which had a box-shaped body with two arms and was capable of lifting a '440-pound barbell', attracting a large crowd that 'waited for as long as three hours' (Haberman 1985). Other robots were set to perform a particular task, including a 'drawing' robot (Matsushita Group), an 'ice sculpting' robot (Hitachi), and a robot that did tricks with spinning tops (Toshiba)—where one trick involved spinning a top on what appeared to be the edge of a samurai sword (NHK Sābisu Sentā [NHK Service Centre] 1985).

Robots were also used as 'actual' performers in a theatrically designed show. The Fuyo Robot Theatre at the pavilion of Fuyo Group, a major business group in Japan, deployed twenty kinds of remote-controlled, wheel-based robots at its main theatre, which could seat one thousand viewers. The theme of the Fuyo Group's pavilion was 'Tomorrow's Science for the care of people', deploying its robots to emphasise human connection, familiarity, and accessibility. These robots had colourful and cartoon-like appearances, with two large, round eyes like the jolly characters of children's television shows (Hoggett 2011).[7] The round and cute appearance of these robots was designed according to given roles, such as the 'friend' robot, the 'baby' robot, the 'cheerleader' robot, the 'wing' robot, and the 'flag' robot. The robot show had robots in pastel colours rolling across the stage in ways that signified cuteness, scooting around or circling, or roaming around each other, blinking large round lights on their heads. Above the stage, in neon, were words in English, 'Robot Fantasy 2001', referencing the science-fiction film *2001: A Space Odyssey* (1968). Around the stage in various scenes during the show were women dressed in pastel-coloured space costumes, looking on benignly or dancing.

Collaborating with Sumitomo Electric Industries, Ichiro Kato presented WASUBOT (WAseda SUmitomo roBOT), a much more sophisticated performing robot that was an outcome of Kato's humanoid research (Sena 2004, 509). This human-sized anthropomorphic robot had a head that looked like a video camera and a torso with two arms and legs that consisted of visible metal-bone structures, wires, and gears. Based on an earlier version, Wabot-2, which was able to read a music score and play an electric organ, WASUBOT performed Bach's *Air on the G String* at the opening ceremony of the Exposition in front of Prince Akihito (the current emperor).

Particularly since the Tsukuba Exposition, robot entertainment shows have become standard exhibits at large, international expositions. The showcasing of the three types of performing robot I discuss for Tsukuba—the industrial robot demonstration, the robot variety show, and the presentation of humanoids of more 'serious' purpose—has become a common promotional strategy taken by corporations for their pavilions.

NATURE AND TECHNOLOGY: GARDEN AND GREENERY EXPOSITION, OSAKA, 1990

The International Garden and Greenery Exposition in Osaka in 1990 was held just before the Japanese economy took a sharp downturn in the early 1990s. Robots intended for scientific work with plants were featured, consistent with an implicit, guiding desire to showcase robotics even where the stated context was to do with horticulture (Fujita and Kinase 1991, 240). Interestingly, despite being a horticultural exposition, performing robots also participated in it through sponsoring corporations. For example, the Mitsui-Toshiba joint pavilion presented a short performance of six one-armed industrial robots, designed and directed by American theme-park entertainment specialist Bob Rogers & Company. These industrial machines were choreographed with music, sound effects, and the projection of animations that created an illusion that the animated pictures were coming out of the ends of the robots' arms. There was one large-sized 'teacher' robot and five smaller 'student' robots (Hanano Banpaku Kenbutsu Gaido Henshū Iinkai [The Editing Committee of the Guide Book for the Flower Expo] 1990, 58–9). They performed a non-verbal music-and-movement-based skit following a humorous storyline in which the student robots did not like the music choices of the teacher robot, but in the end, they performed a new piece of music together. The performance served to anthropomorphise the industrial robots. The Fuyo Group pavilion also presented a short musical theatre performance, with both human actor/singers and remote-controlled robot performers with human voices. The robot performers played robot characters in the performance. Under theatre director Haruhiko Miyajima, the work was a love story between a king of the machine kingdom and a flower fairy (Hanano Banpaku

Kenbutsu Gaido Henshū Iinkai [The Editing Committee of the Guide Book for the Flower Expo] 1990, 52–3). In the story, the technology-centric king realises the importance of nature because of the fairy, and builds a flower garden in his machine dominion. The theme of humanising technology, and the robot, was obvious, as was the idea that 'nature' and technology could coexist harmoniously.

The fact that scientific robots were included in the exposition suggests the presumption that robots should be present. Indeed, the featuring of performance robots at a smaller show such as this indicated a widespread use of the performing robot for expositions, almost as a convention, perhaps because the exposition's sponsors knew that robots would draw a crowd. They evidently saw no contradiction in referencing new technology at a horticultural show, signalling a desire for a harmonious relationship between humans, technology, and nature in twenty-first-century Japan, as the flower fairy narrative demonstrated—facilitated, of course, through corporate knowledge and resources. The figure of the robot stood at the centre of this relation.

In contrast, the unveiling of sophisticated humanoid robotics in exhibitions and industrial fairs from the mid-1990s to the early 2000s presented a demonstration format that focused on technological marvels themselves. In the early 1990s, the 'bubble' burst and Japan's economy plummeted. Japan entered a period of recession and economic stagnation sometimes called the 'lost two decades'. Despite the recession in Japan, new types of non-industrial robots were developed in the 1990s. Honda secretly continued developing a bipedal humanoid from 1986 and, in 1996, unveiled P2, a successful prototype of an autonomous bipedal humanoid. As I have discussed, prominent next-generation robots such as ASIMO, AIBO, and HRP were introduced in the early twenty-first century. In alignment with a persisting theme of harmony and a related model of future cohabitation for humans and robots, the next-generation robots featured at Japanese robot events throughout the 2000s such as Robodex (2000, 2002, 2003), RoboFesta 2001, the Great Robot Exhibition (2007), Japan Robot Festival (2009), and biennial iREX (International Robot Exhibition) in odd years, as well as at Japan's biggest consumer electronic exhibition CEATEC (Combined Exhibition of Advanced Technologies) in 2009, to name a few such notable events. Humanoid robots also featured at the 2005 World Exposition, Aichi.

INTEGRATION AND FANTASY: EXPO 2005 IN AICHI

The main theme for Expo 2005 in Aichi was 'nature's wisdom'—with subthemes of 'nature's matrix', 'art of life', and 'development for eco-communities'—advocating for a better relationship between nature and humans. It aimed to raise ecological concerns for global communities in the twenty-first century through critical reflection on the technological advancements of the twentieth century (Nihon Kokusai Hakurankai Kyōkai [The Japan International Exposition Committee] 2005, 2). As such, environmentally friendly technologies were highlighted and special care was taken to protect the surrounding environment of the site (Nihon Kokusai Hakurankai Kyōkai [The Japan International Exposition Committee] 2005, 6–7). Nevertheless, as in previous Japanese-held expositions that displayed new Japanese technologies of their period, a wide variety of next-generation robots was presented to showcase this important area of Japanese technological development in the twenty-first century.

The Aichi exposition provided a site for the new robots to be shown in use in 'real' situations, in close proximity to visitors. For example, Actroid, a lifelike female android by Kokoro Company, appeared as a human attendant, providing information to visitors when asked questions at information booths.[8] The automated outdoor cleaning robots by Fuji Heavy Industries and Matsushita Electronics were used to clean certain sites overnight (Takenishi 2005, 38 and 44). Visitors were able to interact with NEC's communication robot PaPeRo and the therapeutic robot Paro, a fake-fur-covered robot emulating the size and appearance of a baby harp seal, developed by the state-run National Institute of Advanced Industrial Science and Technology (AIST). (I will discuss Paro in Chap. 8.) Wakamaru, communication robots on wheels, were used as hosts at the pavilion of Mitsubishi Heavy Industries. There were demonstrations of security robots by ALSOK and Tmsuk. The visitor could also experience an automated, intelligent wheel chair, Tao Aicle, by Fujitsu and Aisin. There was a show demonstrating conversation between a human performer and a communication robot, ifbot by Brother. A special 10-day event was organised to exhibit and present demonstrations of a total of 65 prototype robots being developed in Japan (Takenishi 2005, 67). Widespread media coverage helped generate exposure of these newer, service-oriented robots to the wider general public in Japan.

The most spectacular presentation of next-generation robots was offered by the Toyota Group's musician robots, CONCERO (CONCErt and

RObot), and futuristic vehicles, the i-unit and the i-foot. As Aichi prefecture is where Toyota is based, the company poured generous funds into the event, and theirs was one of the most popular pavilions, for which people had to queue for hours. CONCERO consisted of three bipedal trumpet player robots of roughly human height, two wheel-based robots playing trombone and horn, and large, wheeled, tuba- and drum-playing robots of about a metre in height (Takenishi 2005, 9). There was also a communication robot on two wheels that acted as a DJ in the performance. The i-unit was a compact 'concept' vehicle for one person, which was, in essence, a four-wheeled chair with an automated running capability. The i-foot, consisting of an open cockpit and two legs, was a 2.3-metre-high bipedal transporting machine for a single passenger. These machines were the main performers in the daily 30-minute high-tech, multimedia spectacular directed by French multimedia producer Yves Pépin (Mishima 2005, 28). The show included video projection onto a 15-metre-high, 360-degree, 135-metre-long screen that encircled the entire theatre and ceiling and also deployed water fountain devices, fire-flame machines, a team of five circus acrobats, twenty-six floor dancers, and four MCs (Mishima 2005, 28). The spectacle was designed to present Toyota's vision of 'the appeal of mobility – achieving the dream and joy of moving freely' (Mishima 2005, inside cover).

The Toyota performance event consisted of two parts: a 'Welcome show' and the 'Main show'. Within the theme of 'free' and 'joyous' mobility, for humans and for robots, both parts of the show portrayed a seamless integration of human and robot performers playfully frolicking in a techno-Edenic wonderland. Part 1 consisted of music presentations by CONCERO robots, while the DJ robot interacted with a human MC. The robots playing wind instruments were able to play 'real' instruments by coordinated 'finger' movement to depress piston valves and by blowing air into the instrument. The percussion robot was able to hit four drums with a stick in each of its two hands, and play a base drum with a felt beater installed at the lower part of the machine, between its right and left wheels. These robots played their instruments in synch, while moving in the performance space in unison, like a marching band. The second part of the performance, the 'Main show', used lighting and sound dramatically. Video images of running people, various kinds of imaginary vehicles, and abstract butterflies, all avatars of free and easy movement, both human and technological, were projected onto the encircling screen. A solo (human) acrobat was lowered from the ceiling (representing the sky) to a large, five-to-six-metre-diameter dome (as half of the earth's hemisphere) in the middle of the large, round stage. The next scene represented the beginning of time, or the creation of the earth, through a projection of flowing, red magma on the domed earth, with lava filling the visual frame. Dancers clad in futuristic, furred costumes evaded the flames, and onto the scene emerged the i-unit vehicle, representing transportation-to-come as a segue to the next projected scene, one of flattened, brightly coloured flowers and greenery surrounding the dancers and the i-unit (Mishima 2005, 6–7). The dancers approached vehicles of the same colour as their costumes, blending in with them, and rode on selected vehicles. Eight i-units on the stage and another set of six on the circular surrounding stage in the audience area moved in synch with the lighting and music. Later, the i-foot entered the stage, representing 'its first step toward a new history of freedom of movement' (Mishima 2005, 10). The climax of the show had the solo acrobat descend from the ceiling once again. Abstracted images of colourful butterflies were projected over the inside of the performance dome, 'celebrating the beginning of a bright, new future society', according to the guidebook (Mishima 2005, 11).

The gigantic, costly spectacle was designed to present a vision of Toyota's wealth overall, and its penetration into quotidian activity (and the importance of its role in human history) through technological advancement, via the now-expected theme of the robot to signal futurity and prosperity. Especially for the local audience from the Aichi prefecture, who would have been supporters of the company, the robots would have had generated a certain pride and excitement. The robots would have seemed auratic in both senses in which Cormac Power uses the term, that is, to do with fame, and in terms of an aura granted to the robots by their enmeshment in an effective and emotive phantasmagoria.

Cultural Essentialism and Dramaturgy: In Tokyo and on the Australian Tour, 2007

The next-generation robots at Expo 2005 were showcased in accordance with the objectives set by the government-funded Humanoid Robotics Project (1998–2003), which I mentioned earlier; it galvanised the Japanese robot industry to develop newer types of the robot that could work alongside humans. The Humanoid Robotics Project's aims were discussed in terms of the development of new needs and markets for the robot outside manufacturing industries, and that those new markets would be essential for the Japanese robot industry to grow. While Japanese manufacturers of industrial robots were dominant in the world market since the 1980s, their sales growth had been flat for some time due to lack of increased demand for industrial robots (National Institute of Advanced Industrial Science and Technology 2003). Japan's position was in jeopardy due to competition from companies based in the USA, Germany, China, and South Korea. In addition, it was predicted that Japan would face severe labour shortages due to its population decreasing and its ageing population. A 2009 promotional publication of the New Energy and Industrial Technology Development Organisation (NEDO), the government organisation to which I referred earlier that aims to propel the development of next-generation robots, argues in detail for a possible future scenario where robots, providing various services, become part of everyday life at home, in shopping complexes, in the office, or in hospitals (NEDO Books henshū iinkai [NEDO Book Editing Committee] 2009). And in factories, it predicts next-generation robots working side-by-side with human labourers. NEDO also supports the robotisation of agriculture and fishery industries and the development of rescue robots and assist robots capable of working in extreme conditions, for example, at a nuclear power station or in space. There have been numerous book-length publications that support NEDO's views, discussing the outcomes and prognoses for such projects (Nagata 2005; Umetani 2005; Nakayama 2006; Yamato 2006; to name a few).

A common undercurrent of these writings by government sources and proponents of next-generation robotics is a sense of crisis about Japan's future. They look for pragmatic solutions to the falling market for industrial robots and see possible markets in the integration of robots in quotidian contexts. Corporations were already being urged by these government initiatives to think along these lines, and the Aichi expo envisioned them in technicolour for the Japanese public. Two related questions arose: If the robot industry had become the shining symbol of Japanese technological success, under what circumstances could it continue to serve as this symbol and fulfil such expectations in reality? In Japan, with a long history of exceptionalism and a recent imperial past, this question would necessarily become one of Japan's uniqueness, and, at a time of economic crisis after great economic success, its destiny: in what ways could Japan remain exceptional economically and culturally? Discussions of Japan's robotised future became entwined with the narrative of Japanese essentialism.

Two books published in recent years under the same title, which can be translated as *Robots will save Japan*, one by Shin Nakayama (2006), president of Yasukawa Denki, a large robot manufacturing company, and the other by journalist Nobuhito Kishi (2011) after the Great Eastern Japan Earthquake, are highly indicative of this essentialist tendency in relation to the theme of robotics and the pressing question of Japan's future. The title *Robots will save Japan* is used as a slogan in these texts to advocate and argue for the social and economic benefits of the development of next-generation robots in Japan. Techno-nationalistic undertones that have also typified international expos are apparent in both books, and both writers refer to a generalised Japanese affinity with the robot as the background to the success of the Japanese robot industry. With reference to *karakuri ningyō*, Gakutensoku, Astro Boy, the industrial robot boom of the 1970s, the development of next-generation robots, and the Japanese tradition of anthropomorphism, Nakayama bluntly states that it is believed that 'we, the Japanese, have a special feeling toward the robot' (2006, 16).

The Great Robot Exhibition, an exhibition held in 2007 at the National Museum of Nature and Science in Tokyo to celebrate its 130th anniversary, was organised in a way to represent this 'special feeling toward the robot'. In fact, it conveys the message that robots are not simply the subjects of special feeling, as artist Takashi Murakami states: 'Humans regard robots as extensions of themselves and alter-egos. For the Japanese, in particular, robots are the avant-garde of self-portraiture, poised to become reality' (2005, 144, original English).[9]

The introductory statement in the exhibition catalogue is full of terms and expressions that aim to highlight the special position of the robot within Japanese culture. Through the exhibition theme of the world-renowned 'Japanese robot', the producers of the exhibition hoped to tell the longer story of the technological future of Japan, from the past to the present and beyond (Kokuritsu Kagaku Hakubutsukan, Yomiuri Shimbun, and Nihon Terebi Hōsōmō [The National Science Museum, the Yomiuri Shimbun Publishing Company, and the Nippon Television Network Corporation] 2007, 4). In the opening paragraph, robotics as an integration of varied technologies is seen to contribute to Japan's *monozukuri* tradition, a particular view of engineering in which craftsmanship, or the art of making, is appreciated, and is expected to further Japan's advancement. The catalogue statement presents a circular, essentialist answer to the question of why the Japanese love the robot so much, drawing upon the culture of *karakuri*, manga, and animation. According to the rhetoric of the exhibition, for the Japanese, the robot is a dream and something that is to be appreciated and loved, and there is no other country except Japan where people are so fond of the robot (Kokuritsu Kagaku Hakubutsukan, Yomiuri Shimbun, and Nihon Terebi Hōsōmō [The National Science Museum, the Yomiuri Shimbun Publishing Company, and the Nippon Television Network Corporation] 2007, 4).

The first thing the audience saw at the exhibition was a banner that read 'We (the Japanese) love the robot.' This phrase, for Kazuyoshi Suzuki, the curator of the exhibition, is meant to encourage the Japanese to rethink their relationship with the robot, which is cultural as well as technological, along animist lines (Moriyama 2007). The exhibition included *Pan to Rikusho*, an installation piece by film and animation director Mamoru Oshii that featured a large doll figure originally displayed at Expo 2005. Oshii's work was installed at the centre stage of the exhibition as a symbolic representation of unity among earth, nature, and humans. Oshii meant 'Pan', the large doll figure of the installation, to refer to a shaman that brings together animate and inanimate beings and things on the earth and in the universe. 'Rikushō' is a six-part character invented by Oshii to represent the spirit of nature inhabiting the forms of animals. The tripartite integration of nature, human, and robot was regarded as the 'ultimate robotisation' (K. Suzuki 2007a, 10). For Suzuki, Japanese culture is uniquely placed to facilitate unconscious communication between nature, humanity, and the robot, which will lead to a harmonious relationship among them (2007a, 7).

The exhibition consisted of zones with the themes of *karakuri*, manga/anime, and the robot. The *karakuri* section included a display of rare *karakuri* dolls, demonstrations of *karakuri*, as well as a video of a variety of *karakuri* performances, including *dashi karakuri* that is performed on floats in a festival parade. The manga/anime section had a display of tin toy robots, which were important to the Japanese export market in the mid-twentieth century. A large wall cabinet contained hundreds of plastic models of robot anime characters, popular toy products. There was a human-sized, robot character doll as well as a section dedicated to *Astro Boy* and Osamu Tezuka. As well, many of the next-generation robots that participated in the Aichi 2005 exposition were displayed. These included ASIMO, Wakamaru, HRP-2, Paro, and PaPeRo, to name a few. The visitors were able to interact with some of the robots. There were demonstrations of Kondo Kagaku's KHR-2 HV that plays soccer. Twenty universities and colleges also participated, showing the robots they developed: for example, Waseda University's Wabot series robot was presented. Two of Fanuc's industrial robots were also displayed: M-16iA, a handling robot, capable of identifying and grabbing a target object from other objects in a box and transferring it to another box; and M-430iA, designed to pick a target object on an assembly line at high speed. These machines were set up to performing these tasks endlessly.

The exhibition finished at the second exhibition space, which was dedicated to the development of ASIMO. P2 and P3, the prototype models leading to ASIMO, were displayed with a large video panel to illustrate the ASIMO project. At the very end, visitors saw a 15-minute solo skit by ASIMO. The setting for the skit was a domestic situation in the near future, where ASIMO is living with an ordinary nuclear family, parents with a son and a daughter, who appear on a video projection. The plot aimed to show how ASIMO could participate in the everyday activities of family life. The narrator for the show, whom we do not see, explains the robot's functionality, and suggests possible scenarios for the robot in a family setting, indicating that the ASIMO of the future would be able to perform more domestic tasks. For example, in the skit, ASIMO receives a message from the son, who wants to play soccer with ASIMO. The robot starts practicing kicking a soccer ball on stage. The daughter asks ASIMO to help her with her dance lessons. The robot watches the required movements on a video file sent by the daughter. After watching the video, ASIMO replicates the dance movements, while facing the audience. The mother asks ASIMO to prepare drinks for her guests. The robot goes into an entrance at the side of the stage and reappears with four drinks on a tray. It walks across the stage and puts the tray on the table at stage left. The father asks ASIMO to bring his car key to the parking space of the apartment building. The robot walks to stage left, suggesting that he is walking outside the home. ASIMO realises that he has taken the wrong path. The robot then turns 180 degrees, and starts running to stage right. The light fades, indicating the end of the show. ASIMO walks off the stage.

Through the course of the exhibition, the audience of the ASIMO show is encouraged to think that having a domestic service robot is a natural development in the Japanese context, and that that future is just around the corner. In the context of the exhibition, ASIMO acquires a fictional mode of presence, a concept of Cormac Power's that refers to the importance of narrative context to create a character's sense of presence (Power 2008, 15). With regard to ASIMO's performance, it is not only its central role as a character in the skit that renders ASIMO important, a figure to whom the audience is attracted. It is the framing of the exhibition—and, importantly, its situating of the robot as important not only economically for Japan but emotionally—that lends ASIMO its sense of gravity in the Japanese context.

The skit highlights four distinguishable actions that ASIMO is capable of: kicking a soccer ball, dancing, carrying a tray, and running. The audience claps after each action, appreciating the achievement of these movement capabilities by the humanoid robot. Interestingly, in Sydney, the same four actions were presented in a totally different way, and the context for ASIMO's performance signified very differently.

ASIMO's Alive & Unplugged Australian Tour in 2007 toured for a three-month season in major Australian cities. The 25-minute performance was hosted by a professional MC, Simon Miller, at a marquee in the centre of each city. The stage consisted of a performance area with a wall of 16 video screens and two large, electronic lighting boards at the back of the stage. The show started with loud music and flashing lights and rainbow-coloured moving waves on the electronic board, all in synch with the music. After this build-up, the video screen split in the middle and ASIMO walked in. He greeted the audience in English. Speaking with the MC in a casual tone, ASIMO explained what he can do, and why he is in Australia. The robot referred to the Australian tour as part of a worldwide tour to showcase new dance routines by him. After showing some 'stretching' movements, ASIMO walked around the stage. While walking, ASIMO explained what kind of technologies it is equipped with, referring to Honda's cars. The MC asked ASIMO to demonstrate different walking and running speeds. The MC gave the cue for the audience to clap each time. Discussion moved to the technological advancements of ASIMO and Honda's automotive technology, referring to ASIMO's previous models for some background. Inviting a young female audience member to the stage, ASIMO showed his interactive capabilities and demonstrated the action of carrying a tray, while making boyish jokes with MC Miller. Miller asked ASIMO to kick a soccer ball, setting up an imaginary situation that ASIMO is in a football match representing Australia. Their staged conversation returned to Honda's technologies before the final dance section. ASIMO invited a boy and a girl from the audience to dance with him alongside two professional dancers to loud, funky pop music. ASIMO thanked the audience, saying, 'Thanks for having me. It's been totally awesome. You guys rock. See you next time.' When the video screens at the back of the stage opened in the middle, the MC and ASIMO walked off the stage. They turned to face the audience and walked backward while waving to the audience.[10]

These two contexts in which ASIMO was presented, in Tokyo and in Australia, were very different dramaturgically. While the same four actions performed by ASIMO are highlighted in both shows, they framed and contextualised these actions in different ways. While kicking a soccer ball, dancing, carrying a tray, and running were integrated in the storyline for the Tokyo show, these actions were presented in Sydney as simply examples of ASIMO's movement capabilities. The Sydney show focused on Honda's technological advancements, using ASIMO as a 'cool' embodiment of those technologies. It was a promotional event for Honda, elevating its corporate image, with ASIMO as a special ambassador. It is possible to speculate that Honda was careful in its presentation of ASIMO to this Western audience, just as it was in the well-known episode in which Honda sent a delegate to the Vatican in 1996 to consult the Pope regarding the company's development of a bipedal robot. The company wanted to avoid being seen to play God in terms of the Western perception of its robot development work (Hornyak 2006, 109). Aware of the Western emphasis on empiricism and its concern with the rational, Honda may have wished to focus exclusively on ASIMO as a high-functioning machine. In contrast, the show in Tokyo was quite suggestive of future scenarios that touched upon what is understood as the affective relation between the Japanese audience and robots, the same relation that Japan's post-war history of expos and corporate exhibitions has relied upon. For the Japanese, it would be all right, even expected, to identify emotion, even to discuss spirit or 'soul', in a robot demonstration, and certainly to show the robot as a product of its relationality—not as a sophisticated toy or simply a piece of equipment, an object in isolation.

It is possible to speculate that the producer of the ASIMO show, when it was staged at the Great Robot Exhibition, was responding to a newly announced government program, *Innovation 25*, led by Prime Minister Shinzo Abe in 2007. In this program, a clearer government statement concerning the use and purpose of next-generation robots was articulated: that they are to be utilised in concrete domestic and institutional situations by 2025. Government documents show images of possible uses of robots in aged care and in home care (Cabinet Office, Government of Japan 2007). The government website illustrates an imaginary scenario of an ordinary Japanese family, and describes their daily activities with a family robot (Cabinet Office, Government of Japan 2007). The producer of the ASIMO show presented these ideas as a little story, heightening a fictional mode of presence for ASIMO. Moreover, ASIMO's dance in its Tokyo showing was meant to be a '*para para*' dance that features particular arm movements, a commonly known popular dance style from the late 1980s in Japan. For a Japanese audience viewing ASIMO in 2007, the robot's 1980s dance moves would certainly have been familiar and even heart-warming.[11] For the Sydney show, the music chosen was simply loud, with a heavy beat: music for a generalised enthusiasm (or even lukewarm public interest), not crafted to create a culturally specific affect that blended affection, hope, and patriotism in the Japanese context.

Robots, particularly those anthropomorphic robots positioned for operation in the context of human contact in the future, may be seen in Japan as residing within a socially accepted imaginary space unique to them, between beings and things. And it is this imaginary space that is spectacularly actualised in the physical spaces of international and national robot fairs. Just as the *on'nagata* actor's Kabuki performance relies heavily on contextual clues, so does the performative reception of robots by the Japanese audience rely on the careful framing of the robot event.[12] It is for this reason that, at least at this stage, while it is regarded as one of Japan's top priorities, robot development is inextricably intertwined with, and mediated by, Japanese corporate and government aims concerning national identity, consumer acceptance, technology, and futurism. These themes (and pragmatic considerations) converge at large-scale robot events. Such events provide appropriate *mises en scène* for the Japanese next-generation humanoid robots' performance, giving rise to a sense of overt national pride for the Japanese audience. In short, humanoid robot entertainment performances reveal a complex transformative apparatus highly specific to the cultural and historical factors establishing the Japanese context of interest in robots, hidden behind the sophisticated and spectacular appearance of high-tech objects in thrilling, human-like motion. Having presence and social weight, robots become objects that are more than objects in a festive space of transformation and celebration that, drawing upon its traditional antecedent, the *matsuri*, is well known to the Japanese and elicits expected, participatory response.

Importantly, participants in games of make-believe are aware of the fictions of the game. Similarly, when viewing next-generation humanoid robots, I would argue that many Japanese robot enthusiasts are conscious of the limitations and imperfections of the current humanoid robots and are well aware that these robots are not the same as their robot manga idols. It is also unreasonable to assume that all Japanese adopt these social, cultural, and historical narratives in the same way, and they may or may not be conscious of them (and therefore possibly critical of them). The receptive Japanese audience, especially children, appreciates the marvels of the humanoids without necessarily thinking of Toyota's or Honda's long-term corporate goals, for example, or the boost these robots give to Japanese national pride. Writing on the audience's appreciation of the spectacle of professional wrestling, Roland Barthes states, 'The public is completely uninterested in knowing whether the [wrestling] contest is rigged or not and rightly so; it abandons itself to the primary virtue of the spectacle…' (1993, 19, my ellipsis). Similarly, I suggest that as an entertainment of the most advanced machine performance, the humanoid shows effectively offer the audience excitement and wonder and tap into deeper feelings tied up with hopes and expectations concerning the Japanese future. Japanese robot fans are, in essence, responding with love to a collective fantasy of something that does not yet exist.

As I have discussed, there is already an anticipated place for the robot in Japanese society. It remains to be seen what kinds of social and emotional bonds may develop if robots become more technologically and socially sophisticated, and therefore more embedded in Japanese society. Oriza Hirata, in collaboration with Hiroshi Ishiguro, created experimental theatre works that raise these concerns in more dramaturgically refined ways than those of robot expos. I will examine their works in the next chapter.

NOTES

1. I use 'theatrical space' to indicate the space in which the effects that arise in performance take place. In contrast, I understand the term 'theatre space' to refer to a theatre as a building, as well as its actual spaces inside where performances are held.
2. The issue included articles on contemporary performances, such as the works of Socìetas Raffaello Sanzio and Robert Lepage.
3. Book-length studies on the early history of the international exhibition in Europe and the USA include those by Rydell (1984), Greenhalgh (1988), and Mattie (1998). Mattie's study covers exhibitions and expos up to the Hanover Expo 2000, including the Expo '70 in Osaka.
4. Philosopher of science and technology Jun'ichi Murata also discusses how popular industrial exhibitions that promoted Western technologies facilitated the development of local capital-goods

industries and the training of skilled traditional artisans, contributing to the success of Meiji Japan's transformation (2003, 162–173).

5. The list of these prominent academics and artists includes anthropologist and ethnologist Tadao Umesao, science fiction writer Sakyo Komatsu, architect Kenzo Tange, artist Taro Okamoto, as well as younger artists and even less conservative, experimental artists, including architect Kiyonori Kikutake, architect Kisho Kurokawa, stage director Koreya Senda, film director Kon Ishikawa, composer Toru Takemitsu, composer Toshi Ichiyanagi, to name a few.

6. Sawaragi critiques the 1970 exposition in Osaka as a collusion of arts and nationalist propaganda, similar to the art as propaganda presented to the public during World War II (2005b, 62).

7. These robots were designed by Luigi Colani, a high-profile German industrial designer who designed cars for Fiat, Alfa Romeo, Volkswagen, and BMW, and whose design aesthetic is known for its elegantly curved shapes.

8. Actroid is the product name of android machines by Kokoro Co., which collaborates with Hiroshi Ishiguro. See Chap. 4, which discusses Ishiguro's android research.

9. Murakami's vision for robots, in fact, goes beyond the realm of reality, extending into an imaginative space age of a distant future, when 'we Japanese will have developed the robots to protect us, the philosophies to guide us, and the [friendly robot] characters to comfort us' (2005, 148).

10. When I saw this performance while it was staged in Sydney, there were very few audience members, just a few families and children who had stopped to have a look. The low numbers might have been due to the fact that it was a weekday afternoon. In Japan, there would have been a substantial crowd, or possibly a queue to see ASIMO, even on weekdays.

11. Similarly, in 2005, HRP-2 danced the Aizu Bandaisan odori, a popular folk dance routine.

12. The *on'nagata* actor, a male in Kabuki, performs a stylised femininity through codified gestures rather than through simple visual representations, or through an ironic portrayal, as in 'camp' performance. The experienced and knowledgeable audience of this form of traditional theatre appreciates both the actor's technical skills in interpreting and executing a repertoire of patterned acting styles and forms, passed down from one generation to another, and the actor's

presentation of ostensibly 'ideal' feminine qualities through this formalised system. Because these patterns provide a framework for, and the expectation of, audience affect, the Kabuki audience can appreciate an old, male on'nagata actor playing a young female role without viewing his performance as grotesque or comical.

REFERENCES

Barthes, R. (1993). The world of wrestling. In S. Sontag (Ed.), *A Barthes reader* (pp. 18–30). London: Vintage.

Baudrillard, J. (2009). Subjective discourse or the non-functional system of object. In F. Candlin & R. Guins (Eds.), *The object reader* (pp. 41–63). London/New York: Routledge.

Cabinet Office, Government of Japan. (2007). Inobēshon [Innovation] 25. http://www.cao.go.jp/innovation/index.html. Accessed on 10-15-2015.

Clarke, L. B., Gough, R., & Watt, D. (2007). Opening remarks on a private collection. *Performance Research, 12*(4), 1–3. doi:10.1080/13528160701822544.

Eckersall, P. (2013). *Performativity and event in 1960s Japan: City, body, memory.* Houndmills/Basingstoke/Hampshire/New York: Palgrave Macmillan.

Engelberger, J. F. (1999). Historical perspective and role in automation. In S. Y. Nof (Ed.), *Handbook of industrial robotics* (2nd ed., pp. 1–10). New York: John Wiley & Sons.

Fujita, N., & Kinase, A. (1991). The use of robotics in automated plant propagation. In I. K. Vasil (Ed.), *Scale-up and automation in plant propagation* (pp. 231–244). Cell Culture and Somatic Cell Genetics of Plants, Volume 8. San Diego/New York: Academic Press.

Garner, S. B. (1994). *Bodied spaces: Phenomenology and performance in contemporary drama.* Ithaca: Cornell University Press.

Gell, A. (2009). The technology of enchantment and the enchantment of technology. In F. Candlin & R. Guins (Eds.), *The object reader* (pp. 208–228). London/New York: Routledge.

Greenhalgh, P. (1988). *Ephemeral vistas: The expositions universelles, great exhibitions, and world's fairs, 1851–1939.* Manchester/New York: Manchester University Press.

Haberman, C. (1985, May 5). Japanese see a "Made in Japan" future and feel reassured by that vision. *The New York Times*, sec. World. http://www.nytimes.com/1985/05/05/world/japanese-see-a-made-in-japan-future-and-feel-reassured-by-that-vision.html. Accessed on 10-12-2014.

Hakubutsukan, K. K., Shimbun, Y., & Nihon Terebi Hōsōmō (The National Science Museum, the Yomiuri Shimbun Publishing Company, and the Nippon Television Network Corporation). (2007). Goaisatsu [Greetings]. In K. Suzuki (Ed.), *Dai robotto haku: Karakuri kara anime, saishin robotto made* [The Great

Robot Exhibition: From karakuri to anime to the latest robots] (p. 4), exh. cat. Tokyo: Yumiuri Shimbun.

Hanano Banpaku Kenbutsu Gaido Henshū Iinkai [The Editing Committee of the Guide Book for the Flower Expo]. (1990). *Expo '90 Hana no banpaku watashitachi no kenbutsu gaido* [Expo '90, the Flower Expo, Our Exhibition Guide]. Tokyo: Peppu Shuppan.

Hoggett, R. (2011, December 17). 1985: Marco and the Fuyo Robot Theater Expo'85 – Automax (Japanese). *Cyberneticzoo.com*. http://cyberneticzoo.com/robots/1985-marco-and-the-fuyo-robot-theater-expo85-automax-japanese/. Accessed on 10-07-2014.

Hornyak, T. N. (2006). *Loving the machine: The art and science of Japanese robots* (1st ed.). Tokyo/New York: Kodansha International.

Inoue, H. (2007). *Nihon robotto sensōki* [History of Japan's Robot War] *1939–1945*. Tokyo: NTT Shuppan.

Kishi, N. (2011). *Robotto ga nihon o suku'u* [Robots will save Japan]. Tokyo: Bungē shunjū.

Mattie, E. (1998). *World's fairs*. New York: Princeton Architectural Press.

McAuley, G. (1999). *Space in performance: Making meaning in the theatre*. Ann Arbor: University of Michigan Press.

Mishima, T. (Ed.). (2005). *Toyota Group Pavilion Official Guide Book*. Toyota Group.

Moriyama, K. (2007, October 23). Kokuritsu kagaku hakubutsukan, dai robottohaku wo kaisai: ASIMO kara saishin robotto kara, karakuri, anime made [The opening of the great robot exhibition at the national museum of nature and science: From ASIMO to the latest robots, (from) *karakuri* to anime]. *Robot Watch*. http://robot.watch.impress.co.jp/cda/news/2007/10/23/704.html. Accessed on 01-15-2008.

Moriyama, K. (2009, June 17). Nihon Jidō bunka kenyūjo, Aizawa Jirō seisaku no robotto no fukugen sagyō wo kaishi [The Japan Institute of Juvenile Culture, Commencement of the Reconstruction of Jiro Aizawa's Robots]. *Robot Watch*. http://robot.watch.impress.co.jp/docs/news/20090617_294254.html. Accessed on 11-30-2009.

Murakami, T. (2005). Earth in my window. In T. Murakami (Ed.), *Little boy: The arts of Japan's exploding subculture* (pp. 99–149). New York/New Haven/London: Japan Society and Yale University Press.

Murata, J. (2003). Creativity of technology: An origin of modernity?. In T. J. Misa, P. Brey, & A. Feenberg (Eds.), *Modernity and technology* (pp. 151–177). Cambridge, MA: MIT Press.

Nagata, T. (2005). *Robotto wa ningen ni nareruka* [Can a robot be a human being]. Tokyo: PHP Kenkūsho.

Nakayama, S. (2006). *Robotto ga nihon o suku'u* [Robots will save Japan]. Tokyo: Toyō Keizai Shinpōsha.

National Diet Library. (2010). First National Industrial Exhibition. http://www. ndl.go.jp/exposition/e/s1/naikoku1.html. Accessed on 03-03-2015.

National Institute of Advanced Industrial Science and Technology. (2003). Hataraku ningengata robotto: Ningen kyōchō, kyōzongata robotto sisutemu no kenkyū kaihatsu wo happyō [Announcement on the research concerning the robot system that cooperates and cohabits with humans]. http://www.aist.go. jp/aist_j/press_release/pr2003/pr20030226/pr20030226.html. Accessed on 05-07-2014.

NEDO Books henshū iinkai (NEDO Book Editing Committee). (2009). *RT supirittsu: Hito ni yakudatsu robotto gijutsu o kaihatsu suru* [RT spirits: Developing useful robots for humans]. Kawasaki: New Energy and Industrial Technology Development Organization.

NHK Sābisu Sentā [NHK Service Centre]. (1985). *Nijū isseki ga mieta kagaku banpaku 85* [The 21st century is in sight at expo '85] (VHS). Tokyo: Tōhō.

Nihon Kokusai Hakurankai Kyōkai [The Japan International Exposition Committee]. (2005). *2005 nen nihon kokusai hakurankai kyōkai: Ai chikyūhaku kōshiki gaido bukku* [2005 Japan international exposition committee: Love the earth expo official guide book]. Tokyo: Pia.

Osakafu Nihon Banpaku Kinen Kōen Jimusho (Expo '70 Commemorative Park Office). (2014). Taiyō no tō [The Tower of the Sun]. *Banpaku Kinen Kōen.* http://www.expo70-park.jp. Accessed on 15-10-2014.

Power, C. (2008). *Presence in play a critique of theories of presence in the theatre.* Amsterdam/New York: Rodopi.

Roche, M. (2000). *Mega-events and modernity: Olympics and expos in the growth of global culture.* London: Routledge.

Rydell, R. W. (1984). *All the world's a fair: Visions of empire at American international expositions, 1876–1916.* Chicago: University of Chicago Press.

Sawaragi, N. (2002). Isozaki Arata Intabyu: 1970 nen, Osaka banpaku wo kataru [Arata Isozaki interview: A talk on the 1970 Osaka Expo]. In Kirin Puraza Osaka (Ed.), *Expose 2002* (pp. 10–11). Osaka: Kirinbīru KPO Kirin Puraza Osaka.

Sawaragi, N. (2005). *Sensō to banpaku* [War and expo]. Tokyo: Bijutsu Shuppansha.

Schodt, F. L. (1988). *Inside the robot kingdom: Japan, mechatronics, and the coming robotopia.* Tokyo/New York: Kodansha International.

Sena, H. (2001). *Robotto Seiki.* Tokyo: Bungei Shunjū.

Sena, H. (Ed.). (2004). *Robotto Opera (Robot Opera: An anthology of robot fiction and robot culture, original English title).* Tokyo: Kōbunsha.

Sugano, S. (2011). *Hito ga mita yume, robotto no kita michi: Girisha shinwa kara Atomu, soshite* [Mankind's Dream, The Historical Path: The Robot From (ancient) Greece to Astro Boy]. Tokyo: JIPM Sorūshon.

Suzuki, K. (Ed.). (2007a). *Dai robotto haku: Karakuri kara anime, saishin robotto made* [The Great Robot Exhibition: From Karakuri, to anime, to the latest robots]. Exh. cat. Tokyo: Yumiuri Shimbun.

Takenishi, M. (Ed.). (2005). *Aichi banpaku saishin robotto gaido* [Aichi Expo. The guide to the latest robots]. Bessatsu Robocon Magazine. Tokyo: Ōmusha.

Umetani, Y. (2005). *Robotto no kenkyūsha wa gendai no karakurisi ka?* [Are robot researchers contemporary Karakuri Craftsmen?]. Tokyo: Ōmusha.

Wilson, S. (2012). Exhibiting a new Japan: The Tokyo Olympics of 1964 and Expo '70 in Osaka. *Historical Research, 85*(227), 159–178. doi:10.1111/j.1468-2281. 2010.00568.x.

Yamato, N. (2006). *Robotto to kurasu: Kateiyō robotto saizensen* [Living with a robot: The frontline of domestic robots]. Tokyo: Sofuto Banku Kurieitibu.

Yoshimi, S. (1992). *Hakurankai no seijigaku* [The Politics of Expo]. Tokyo: Chūkō Shinsho.

Yoshimi, S. (2005). *Banpaku gensō: Sengo seiji no jubaku* [Fantasies Through the Exposition: The Curse of Postwar Politics]. Tokyo: Chikuma Shinsho.

Yoshimi, S. (2008). *Toshi no doramatrugī: Tokyo sakariba no rekishi* [Dramaturgy of the city: History of entertainment districts in Tokyo]. Tokyo: Kawade shobō.

CHAPTER 4

The Anthropomorphic Robot and Artistic Expression

In this chapter, I discuss the figure of the robot as a means to dramatise differences between the human and the robot. In addressing this theme, I will examine robot and android theatre works by playwright and director Oriza Hirata that feature roboticist Hiroshi Ishiguro's humanoids and androids, and Ishiguro's android performance experiments outside the spaces of theatre from 2009 to the present. Ishiguro's robots are regarded as next-generation robots, and collaborative works by Hirata and Ishiguro, and Ishiguro's own experiments, have received official support, funded by Japanese government grants. These works will be contrasted with artist Kenji Yanobe's installation works presented in a visual arts context, including his development of a fire-breathing, gigantic robot in 2005. The examples of robot performance I treat in this chapter present a much more complex performance matrix than those in the expositions that I discussed in the previous chapter. Like those robots performing in expos, the robots in works by Hirata, Ishiguro, and Yanobe also call upon affective response in their Japanese audiences, yet there are notable points of difference.

According to Erika Fischer-Lichte, '[an] actor appears to spectators as a sort of magical mirror that reflects their own image back to them, allowing them to see themselves as others' (2014, 184). I discussed in the previous chapter that robots may occupy a special place in the Japanese imaginary—neither human, nor object. Though Fischer-Lichte is speaking of the transformations created by human actors, given the relation robots have to the human, we can also ask if a performing robot 'actor' can be seen, literally or

© The Author(s) 2017
Y. Sone, *Japanese Robot Culture*,
DOI 10.1057/978-1-137-52527-7_4

metaphorically, as an agent that reflects what the human is or performs its proximity to the human.

In Hirata's dramaturgy, theatrical devices to do with dialogue are used to suggest the robots' 'humanness' in imaginary situations (in the near future) of cohabitation by humans and robots. This is, by now, a familiar theme for the Japanese audience since the expos of the early 2000s that promoted it. The robot characters invite empathetic responses from audience members. Hirata's theatrical apparatus works hard to hide the liminality of the robot. At the same time, the robot's very in-between status, and stagecraft that emphasises the robots' human-like qualities, allow his plays to use their robot characters—that is, characters that are acknowledged at the narrative level of the plays as robots—to represent the social 'other'. Ishiguro has indicated more broadly conceived concerns for the explorations of his robot works. Ishiguro discusses his performance androids (both inside and outside the theatre) in relation to the question of humanness. For Ishiguro, the liminality of the android hinders people's motivation to communicate with his machines. Ishiguro's search for a smooth integration of androids in social situations has led to his latest experiments conducted in the form of an android comedy performer in the Japanese popular-culture context.

As a contrasting example, Yanobe's work is a tall, metallic robot statue that was presented in a gallery as art. It is an enlarged version of the fictional character 'Torayan', a ventriloquist doll. This gigantic robot induces a mixed sense of the uncanny and of awe, a departure from the common framework for discussions of the robot in Japan as set by government officials, corporate concerns, and roboticists—and Hirata's and Ishiguro's concern with human-likeness. According to the party-line, next-generation robots are supposed to be familiar and friendly and operate in close proximity to humans. Hirata's approach, to incorporate robot characters in his narratives and render them the subject of concern and empathy, and Ishiguro's interest to find ways to integrate robots socially are both consistent with national and corporate objectives. Yanobe's robotic works, however, suggestive of the transgressive spirit of *matsuri*, appear strange and disturbing. Yanobe's robotic machines bring to the fore the memory of the bleaker, contradictory side of Japanese modernity, science, and technology.

Oriza Hirata's Robot Theatre

Oriza Hirata and Hiroshi Ishiguro have worked together to produce several robot theatre performances: *I, Worker* (2008), *In the Heart of a Forest* (2010), and *Sayonara (Goodbye)* (2010), *Three Sisters, Android version* (2013), and *La Metamorphose Version Androide* (2014). In their collaborative works, three types of Ishiguro's robots are used. Wakamaru and Robovie R3 are humanoids that are able to vocalise pre-recorded talk while their operators control the timing of the robots' actions. Geminoid F and Repliee S1 are female and male androids, respectively, that closely resemble their human models and have lifelike skin, hair, and gestures. They are each operated by a human actor who provides their speech, while the actor's facial movements are transmitted to the robot's face. These androids were integrated as actors into Hirata's plays and rendered believable for the audience through his stagecraft, the plays' settings, lighting, and storyline. I suggest that Hirata presented these robots as friendly and humane characters so that the Japanese audiences for these plays could relate positively to them.

The drama of co-existence between the human and the robot is the theme of Hirata and Ishiguro's robot theatre projects, echoing official narratives promoted by the Japanese government, roboticists, and robot industries, as I have indicated. For collaborative projects from April 2011 to March 2014, Hirata was successful in receiving 49,000,000 yen (approx. AUD $ 510,000) from the prestigious government fund of Grants-in-Aid for Scientific Research from the Ministry of Education, Culture, Sports, Science and Technology (National Institute of Informatics 2011). Hirata states that the most pressing aim of the project was 'to transform a conventional display of robots at scientific expositions into a robot theatre of artwork' (2012, original English). For Hirata, the scientific robot exhibitions do not engage with the viewers emotionally, though they may be intellectually intriguing. Hirata states that the collaborative projects with Ishiguro would provide 'a basis for research into developing future robots that will not make elderly people and children feel uncomfortable or intimidated' (2012, original English).

Hirata and Ishiguro's theatre projects highlight the possibilities for emotional bonds between humans and robots. Hirata sets the scenes of these productions in a near future in which robots are part of the household and its everyday routines, interacting with humans daily and casually. For example, in *I, Worker*, a young married couple has two domestic robots. While the female robot does household chores as it is supposed to do, the male robot has trouble finding any motivation to work. It is a story about a robot facing an existential crisis or retreating entirely as if it is *hikikomori*, a Japanese social phenomenon in which young people, mostly but not always men, become recluses in their homes. *In the Heart of a Forest* is set in the Democratic Republic of Congo and is about a group of scientists researching the differences between humans and bonobo apes. Robots act as assistants to the researchers. The play explores the boundaries between the three so-called 'species'. *Sayonara* is about a young woman with a fatal illness. Her parents buy her a robot companion and then leave her alone with it. *Three Sisters, Android Version,* an adaptation of the play *The Three Sisters* (1900) by Anton Chekhov, depicts three sisters' lives in a declining regional city where once its robotics industry boomed. The family lives with a devoted domestic humanoid, as well as an android that was supposedly a substitute for the dead youngest sister, Ikumi. The story reveals that Ikumi was in fact not dead, but *hikikomori*. The android was used by Ikumi as an avatar. Hirata's stories reflect contemporary social issues in Japan such as isolation, loneliness, or apathy through relationship dramas involving humans and their companion robots. *La Metamorphose Version Androide,* based on Franz Kafka's 1915 story 'The Metamorphosis', was performed by French actors in French, with projected Japanese subtitles, for its seasons in Japan. In Hirata and Ishiguro's version, Gregor Samsa, the main character, transforms into an android instead of the insect-like creature of the original story. While Gregor does not accept his fate, his family eventually becomes used to it; at the end of the story, the family refuses to turn off the power switch when he asks them to do so because he wants to die. The overall intention of these plays is that these robot characters assist in essential ways to reveal, through the robot characters' interactions and relationships with the human characters, the complexities and anguish of human existence.

It is important to highlight that Hirata's dramaturgical philosophy and strategies are developed from themes in his past work and are reflective of the Japanese context. Hirata is known as the advocate of 'contemporary colloquial theatre' (*gendai kogo engeki*), which was thought to have ignited the emergence of 'quiet theatre' (*sizukana engeki*) in the 1990s. The latter demonstrates a psychological realism that shows, in a meticulously detailed manner, the vocal and gestural mannerisms of contemporary Japanese people, in contrast to the busy, exaggerated gestures and excessive vocal utterances of 1980s theatre practises in Japan, as in the work of Hideki Noda or Shoji Kokami (Katsuya Matsumoto 2015, 7). Hirata's plays depict details of Japanese social relations, highlighting 'an induced capacity for social forgetting, a profound cultural amnesia and form of advanced alienation' (Varney et al. 2013, 71). Hirata's robot theatre project is intended to show how Japanese people might engage in a dialogue with a robot as if it were '*kūki ga yomenai hito*' (a person who cannot feel out the situation, often described colloquially in Japan as 'KY'), in other words, as an unpredictable outsider or social other (Ishiguro and Oriza 2015, 123). Using robot characters that express their existential anguish, not dissimilar in essence to Astro Boy's dilemma, Hirata taps into contemporary issues of otherness, aloneness, and interpersonal communication problems that are prevalent in Japan.

The collaborative robot theatre projects of Ishiguro and Hirata present a form of contemporary storytelling using robots where puppets might otherwise have been used. In the Ishiguro-Hirata collaborations, performance techniques similar to those used in Bunraku were mobilised to heighten the sense of communication between the human and robot characters. For example, according to Kazunari Kuroki, the producer of *I, Worker*, a Bunraku master (Kanjuro Karatake) and a pantomime artist (Naoki Iimuro) were invited to provide advice on gendered gestures (2010, 62). Hirata was also very attentive to speech 'gestures'. In his attempt to conjure convincing characters from the robot 'actors', Hirata focused on the timing of verbal exchange between the robot and human characters and the timing of the humanoids' enunciations was calculated and adjusted to a tolerance of less than a second (Hirata et al. 2010, 18). Hirata directs his (human and robot) performers at the limit of human cognition.[1] There were also strategic uses of interjectory and exclamatory phrases such as '*eh?*' or '*ē, mā*' that are used frequently and characterise Japanese oral communication. They are modulators in Japanese speech, making the speaker's words softer or stronger as well as direct or indirect. Replicating the subtleties of speech would convey to the audience the signs of communication. Such interjectory phrases, expressed through melodic inflexions, occur in speech as in-between words and breath, utterance and non-utterance. These expressions highlight literary theorist Steven Connor's pneumatic concept of voice as 'the sounding of hollowness', pointing to a relationship among voice, breath, and vital force (2000, 35). These sounds would not only offer the signs of communication but mimic breath and thereby signal life for the robot characters.

As a dramatist who has paid detailed attention to Japanese language and its utterance, Hirata knows how the Japanese express their emotional sentiments in conversational forms. Concerning the 'speech' and gestures of robot characters, Hirata pays attention to '*ma*' (meaning interval), sometimes includes an inversion of the normal order of words, and considers gestural mannerisms and regional accents. Because Hirata's use of interjectory and exclamatory phrases and other conversational devices was very effective in these robot plays, a sense of human interiority was projected on these human-looking machines for the Japanese audience, which was already familiar with human-like robot characters appearing in manga/anime. What Hirata's approach highlights is the effectiveness of contextual cues in a dramatic setting (coupled with facial and physical cues performed with absolutely precise timing) to suggest a living presence for the Japanese audience willing to engage with the play's theatrical premises.[2]

The robot projects are thought to be a continuation of Hirata's theatrical experimentation since the mid-1980s, which considers, according to critic Atsushi Sasaki, the viewer's sensing of what is '*riaru*' (real) through the fictiveness of theatrical drama (2011, 286 and 325). Theatre studies scholar Kei Hibino describes Hirata's dramaturgy as an oscillation 'between fakery and authenticity' (2012, 30, original English). In this sense, Hirata's approach is a perfect match for the social staging of the robot in Japan, reflecting both the wish to be pleasurably fooled evident in manga/anime and in the robot entertainments of earlier expos, and more recent themes of cohabitation and 'real' interaction, which have been cultivated through the robot's representations at large-scale exhibitions since the 2000s. Hirata believes that theatre can provide a 'context' for viewers, offering a virtual experience of other people's lives through his stage characters. For Hirata, an understanding of a new and different perspective experienced from theatre art, and the sharing of that knowledge, generates a sense of the real (1998, 202–3).

Hirata holds the controversial view of acting that an actor does not need to draw upon his or her interiority to inform the character.[3] Interiority is also known in Japanese terms as *kokoro*, meaning the soul, the spirit, the mind, or consciousness. The theatre audience for traditional realism—as in Stanislavski's theory, or in method acting—is encouraged to believe that the emotional expression of a character in a play matches the actor's internal state on stage. Hirata rejects this idea and stresses that play-writing should be precise and robust enough so that it can be performed correctly without an actor's 'unique' expression, as such (1995, 67–8). For Hirata, a play should be written in the most detailed manner possible, including the timing of the utterance of actors' lines. It is the director's job to transpose the fictional world of a play to the stage, coordinating actors and other theatrical devices. The actor's task is to trace what is written in the play, and to present his or her role as faithfully to the writing as is possible. It should not have to be about expressions that correspond to an actor's internal feelings. Hirata rejects the criticism from the perspective of realist theatre that his approach renders acting superficial and even regards the 'superficiality' as justified (2004, 185). All that matters for Hirata is that the actor convinces the audience, and a well-controlled robot can certainly do just that.[4]

According to Kuroki, a questionnaire completed by the audience of *I, Worker* suggests that nearly three-quarters of the audience felt that the characters played by humanoids demonstrated a sense of human interiority on the stage (2010, 65). Hirata felt vindicated: he was able to show evidence that an actor does not need to demonstrate his or her *kokoro* as long as the character is perceived to have it; and, so, a robot is good enough as an actor (Sasaki 2011, 310). Hirata asserts that half of theatre is created in 'the brain' of the audience, meaning that the viewers' active interpretation of the acting, stagecraft, and totality of the theatrical experience is essential to the meaning of theatre works (Hirata et al. 2010, 20). For those who regarded the production favourably, the robot performers on stage were perceived in a positive light because the robots were interesting as robots (Cormac Power's literal mode of presence), they were believably part of the narrative (the fictional mode of presence), and their characters seemed to have *kokoro*. We can interpret this last aspect in terms of Power's version of auratic presence in which the overall context creates a charismatic effect around the characters: we know the context is necessarily important precisely because of Hirata's careful work to create the impression of liveliness, of a life force inhabiting the humanoids.

Ishiguro, the technical director behind these robot and android theatre projects, holds a similar view on *kokoro*, though approaching the issue of *kokoro* from a different angle. I will discuss the differences in approach between Hirata and Ishiguro by examining the android theatre project *Sayonara* that featured Geminoid F, a female android. *Sayonara* received a mixed reception. I will first briefly discuss the background of Ishiguro's Geminoid project before discussing *Sayonara*.

The Failure of *Sayonara*

Through his earlier experiments with humanoids that could communicate with humans verbally with pre-recorded responses, Ishiguro came to conclude that robots that have a human-like appearance can more easily be anthropomorphised and therefore people might find it comfortable to communicate with them. For Ishiguro, *mikake* (appearance) and movement of interactive robots are just as important as technological innovation and improvements that are less immediately noticeable (2007, 185). For these reasons, he decided to experiment with androids that resemble humans in the early 2000s. Repliee Q1expo was an automated conversation android that was exhibited at Expo 2005 under the general name Actroid, in collaboration with Kokoro Co. However, in order to make a robot that could sustain a conversation with a human using the current, limiting technologies, Ishiguro built an android that could be operated remotely (2007, 261).[5] The operator would be stationed in a different location from the android and would speak through a microphone. The android would function as an avatar while replicating the involuntary bodily movements of the operator's eyes, head, chest and shoulders involved in breathing. Ishiguro also decided to use himself as a model for the android this time and named the android he developed in 2005 'Geminoid HI-1'. Geminoid F, created in 2010, is the second Geminoid, modelled on a woman with a Eurasian-appearance whose identity was not revealed. Based upon his experience with Geminoid HI-1, Ishiguro realised that the android with his own appearance—which has a limited range of expression and a very serious countenance—was limited in its capacity to communicate with children (2012, 95). Ishiguro also wanted Geminoid F to be less ethnically specific so that it could be used for experiments in countries outside Japan (2011, 34). It was designed with a simpler operating system, so it could be more easily transported and used in varied contexts (Ishiguro 2012, 95).

Ishiguro asked Hirata to write a play using Geminoid F, which culminated in *Sayonara* (Ishiguro 2013, 147). This short play is a 15-minute dialogue between a young woman (played by Bryerly Long), who is suffering from an incurable disease, and her android companion (Geminoid F), which, in the storyline, has been provided by the woman's father. Geminoid F is seated on one side of the stage, and Long is seated on a rocking chair. The play is a static piece, with little movement on stage. Geminoid F was able to move her head, eyes, and mouth and also, in a limited way, move her shoulders and chest. When talking, the robotic machine is operated by actor

Minako Inoue off-stage, and Inoue contributes the voice of the android character. In the play's narrative, the woman takes solace in the company of the android as she comes to terms with her imminent death. The android tries its best to comfort her by reciting well-known poems by French poet Arthur Rimbaud, German poet Karl Busse, Japanese poet Shuntaro Tanikawa, and *tanka* (a type of Japanese traditional poetry) poet Bokusui Wakayama. When the android recites Rimbaud's poem in Japanese, the woman replies with Rimbaud's poem in French from her childhood memories. When the human character recites Busse's poem in German, the robot repeats the poem in Japanese. These poems serve to express feelings of apprehension, loneliness, and longing and a sense of closeness and also distance when one is about to embark on a journey to a place far away and unknown. The play contrasts human mortality with the immortality of the android. In 2012, as the second version of the play, Act 2 set was added, which was set in a time after the death of the woman.[6] The android was to be sent to the no man's zone of the Fukushima nuclear disaster to comfort its victims.

Critical reviews highlight the mawkishness of *Sayonara*'s theme and the unnaturalness and stiffness of Geminoid F's limited performance. According to Kei Hibino, the Japanese audience saw the show as 'a kind of technological hocus-pocus that glosses over a plotless drama charged with too much sentimentality' (2012, 36–7, original English). On the English version of the production, which used the recorded voice of Long for the android, the critical reception from non-Japanese writers was mixed. For example, for Australian theatre studies scholar Glenn D'Cruz, it was 'a compelling failure' (2014, 273). For Canadian reviewer J. Kelly Nestruck, because Geminoid F's movements and the recorded voice were marginally out of synch, '[a]s soon as Geminoid F begins to speak and move around jerkily . . . the illusion slips away' and, thus, it was 'a lifeless production', and it was 'pretentious' (2013, my ellipsis). For these critics, Geminoid F was a mere puppet representing a gender stereotype, as its female (android) character was a carer (D'Cruz 2014, 286; Borggreen 2014, 159; Nestruck 2013). Nestruck concluded that Geminoid F does not 'pose any threat to [our] future livelihoods' (2013). For theatre critic Kentaro Mizu'ushi, *Sayonara* is not a theatre piece but a mere spectacle to demonstrate the human-like android (2011). This observation resonates with Parker-Starbuck's ironic comments that 'in *Sayonara*, through Geminoid F, we see a return to the "entertainer"' (2015, 76). These critical comments suggest the misalignment of expectation and delivery in Geminoid F's performance and in terms of the play's other contributing aspects, which did not gel in a way that would, as was the case for *I, Worker*, effectively create a sense of presence around the android character.

The disharmonies within *Sayonara* could have been seen as an opportunity to explore a more complex sense of social otherness with regard to the Geminoid F. Both of the characters of *Sayonara* are not typical Japanese characters: the Eurasian-looking Geminoid F 'speaks' with Inoue's native Japanese, while Long's character speaks three languages (Japanese, French, and German); her heavily accented Japanese would sound peculiar to a Japanese audience. As well, in Japan, one rarely encounters a situation where several European languages are spoken in the same situation. *Sayonara* was imbued with foreign otherness, as well as the unintended unnaturalness of the android, which was highlighted for viewers of this work by its out-of-synch speech and limited range of movement. Theatre critic Osamu Nakanishi suggests that the aim of an android theatre may be to make our invisible and unconscious ways of viewing others more visible (2014). This question could have been explored in *Sayonara*. However, the narrative structure of the play does not allow for a reading of the android that might embrace these structural peculiarities in order to explore its real 'foreignness' (for which the use of multiple European languages may be a metaphor); instead, its signification is fixed within modes of presence related to a theatre context and, more generally, within the terms of the government/corporate narrative of future harmony and cohabitation.

Hiroshi Ishiguro's Android Experiments

From Ishiguro's perspective, *Sayonara* was a great success, much more than his previous robot theatre experiments (2013, 149). His concern was that the android project had progressed his robotics research. Ishiguro was concerned to investigate meaningful communication between humans and robots and felt that the work had established a situation for that exchange. While Hirata's interests in the use of the robot are technical in terms of theatre-making, Ishiguro's interests go beyond those of engineering, intersecting with cognitive science, developmental psychology, and brain science. For Ishiguro, his anthropomorphic machines are tools for the investigation of human communication and interaction, as much as they may be intended, in terms of larger, stated objectives, as prototypes for specified functional purposes in future workplaces. In addition to claims that his research is of direct relevance to robotics design, Ishiguro also regards robots as a mirror of the human when asserting the larger significance of his research. He is not afraid to ask questions and make statements usually reserved for philosophers, such as: 'What is the human?'; 'What is left in the human if all human abilities are replaced by machines?'; 'What is it that creates a human-like quality? How can it be expressed in a robot?'; 'Can a robot have *kokoro*?' (Ishiguro 2007, 2009).

The use of the term '*kokoro*' in robotics research may sound strange, but Ishiguro is not alone in using it. Other Japanese roboticists who discuss the *kokoro* of the robot in their publications include Minoru Asada (2010), Jun'ichi Takeno (2011), and Tadashi Nagata (2005), to name a few. Questions of *kokoro* arise in an area within Japanese robotics that is concerned not only with human intelligence but also with consciousness or mind, including attitude or emotion. A sense of affection toward a robot can be explained in relation to the Japanese animist tradition that in the past resulted in the veneration of tools such as the hoe and sickle, and the sanctification of newly built industrial machinery in contemporary contexts. Philosopher Shozo Omori argues that the animistic projection of *kokoro* is common in Japanese society, and it is a matter of one's attitude toward an inanimate object, including a robot, and how one establishes a long-term relationship with that object through daily activity (1981, 72 and 140).[7] This correlation between animism, empathetic projection, and robots continues to be discussed in recent writings by proponents of next-generation robots, including Ishiguro (Okuno 2002, 122; Ishiguro 2007, 181; Sonoyama 2007, 151; Sena 2008, 415; Asada 2010, 161).[8]

For Japanese roboticists, this area of enquiry concerning a robot's *kokoro* is strongly linked to AI (artificial intelligence) research in the Western context. Debates within Western AI-related studies are noted from the particular perspective of Japanese robotics. Indeed, Ishiguro proposes that his Geminoid could take the Total Turing Test (2009, 127); Ishiguro would add verbal and gestural communication to the Turing Test with the android.[9] Ishiguro and Japanese roboticists' interest in *kokoro* relates to discussions on 'Strong AI', the term coined by philosopher John Searle to critique the assertion that 'the appropriately programmed computer literally has cognitive states and that the programs thereby explain human cognition' (1980, 417). 'Strong AI' is used to describe a certain position within AI research, the aim of which is to build artificial intelligence that is equal to human intellectual capacity. In contrast, 'Weak AI' suggests a more pragmatic approach concerning computer intelligence, limited to the parameters for a computer to simulate only some aspects of human cognitive and physical behaviour, to which Searle does not object (1980, 417). Roboticist Takashi Maeno describes himself as one of those who believe in 'Strong AI' (quoted in Hotta 2008, 99). Ishiguro states that in his pursuing of the robot's *kokoro*, the ultimate goal is 'to create a human' (2007, 186).

Ishiguro approaches the question of the robot's *kokoro* from a specific angle, saying that humans do not have *kokoro* and it is only a belief that humans have it (2009, 3, 2011, 43). Or, at least, it is impossible to prove if that is right or wrong (Ishiguro and Ikeya 2010, 158). Ishiguro holds a mechanistic view of the human body, such that most of its functions can be replaced by machines (2012, 48). For Ishiguro, the human body is a mere sack full of faeces (2013, 22). If so, the notion of the human should be defined in terms of one's ability to form relationships, that is to say exteriorised encounters, with other humans, and, further, that these relationships are based upon mechanistic exchanges built of specific gestures and behaviours that can be replicated (Ishiguro 2012, 49). If one feels the *kokoro* of someone in an exchange, one only thinks so by observing that person's actions, on the basis of a preconceived belief that humans have *kokoro* (2013, 165). Ishiguro therefore states that it is possible for a robot to have *kokoro* if humans are led to sense that the robot has it, as in the production *I, Worker* (2011, 130).

Ishiguro's position echoes the views of other Japanese proponents of the idea that the robot can have *kokoro*. For example, Maeno stresses that it is good enough if a robot is able to make appropriate comments in an appropriate time, even if the comments are not informed by a subjective intention, as such (interviewed by Hotta 2008, 103). For philosopher Masayoshi Shibata, a robot that appears to have *kokoro* should be regarded as a robot with *kokoro* (2001). Both Maeda and Shibata bypass the question of 'intentionality'. Likewise, Ishiguro focuses on the appearance of the robot and the android, that is, external demonstrations, rather than on any kind of subjective motivation, an approach that tallies with Hirata's in regard to actors' performance, as I have discussed earlier.

The collaborative theatre projects with Hirata provide Ishiguro with test 'laboratories' outside ordinary robotics contexts to find effective ways for his robots to engage with people. In the theatre projects, the robot programming is designed specifically for the particular needs of a particular performance project. Ishiguro has learnt from the theatre experiments that by specifying contexts and presentational frames, his robots and androids are able to engage with people in a manner perceived as more meaningful, and, therefore, viewers would be more likely to sense a robot's presence, which the audience of *I, Worker* reported as *kokoro*.

Apart from working with Hirata, Ishiguro carried out performance experiments in non-theatre spaces, which could be seen as a kind of performance art. In 2009, Ishiguro presented an artist's talk using his double, Geminoid HI-1, at Ars Electronica, an international media arts festival held in Linz, Austria. At the end of the piece, the Geminoid simulated 'death' through the effects of breath. It was a dramatisation of the relationship of voice and breath to suggest one's (the android's) vital force. The 'death' of the android was indicated at the moment when air was released from its pneumatic system. There was something uncanny in this scene of the Geminoid's death. Its head fell backward slightly and it became utterly still. Perhaps its machinic strangeness was revealed, in contrast to the avatar/puppet structure that was operative until that point. The dramatic moment of this 'death' emerges as a result of a human projection of an idea of death onto an inanimate machine. It is an anthropomorphic performance of boundary transgression, where air signifies the vital force of human breath. However, the difference of the robot's 'death', the peculiarity of its 'exhalation' and utter stillness, opened an opportunity for the irreducible difference of the robot, its utter non-human-ness, to be explored, an opportunity that was not taken. Ishiguro also conducted an experiment with Geminoid HI-1, placing it at a café in Linz and speaking through it to test how people (who knew the 'installation' was a part of Ars Electronica) would react to the machine in a social situation.

Ishiguro seeks ways of using his robots in real social situations. Ishiguro's Actroid robots, automated communication machines, were used as guides at information booths or as MCs at pavilions at the Expo 2005 in Aichi. More recently, for one of the Geminoid F events in 2012, which was designed for Valentine's Day, Ishiguro created a scenario in which a woman (the android) is waiting for a lover. Geminoid F was programmed to perform 'anxious waiting' through its facial and behavioural expressions at a display window that looked onto the street at a Takashimaya department store in Tokyo. At an Osaka branch of the same department store, Geminoid F was given the role of sales assistant. A visitor could ask questions via preloaded comments on touch screens. The female android answered with pre-recorded messages. At these performance events, Ishiguro hoped to provide social 'frames' for viewers as a function of the Geminoid's specific role, within which the android becomes meaningful. While Ishiguro's androids are not immediately ready to be deployed in actual situations, these social experiments provided him with wide media exposure of his android research.

While androids experimentally placed in service roles within social situations were received reasonably well by the public, Ishiguro wished to explore how reception, and the perceived aura of the robot, could be enhanced by locating androids in Japanese popular media contexts. In essence, Ishiguro capitalised on the gloss a person acquires (Power's concept of aura, through fame) by exposure in the media. Even better, would a robot that looked like a famous person become equally famous? Ishiguro started producing robotic 'wax dummies' of famous people. In 2012, collaborating with make-up artist Shinya Endo, Ishiguro created an android replica of a well-known *rakugoka* (classic comic storyteller) Katsura Beicho III (who died in March 2015), who was designated a national living treasure in 1996. Arts production company Breeze Arts in Osaka approached Ishiguro to build the android in order to record and preserve the tradition of '*kamigata rakugo*' (Kansai-based classical storytelling from Osaka and Kyoto). Beicho Android performs by synchronising his facial and gestural movements with the recorded voice of Beicho. The aura of Beicho and the comedic aspect of *rakugoka* helped to alleviate the technological limitations of the android. (In Chap. 7, I discuss the use of humour in comedy robot competitions, with similar effects.)

As a strategy to mask technical issues, Ishiguro further developed the idea of the comedic android in Real Android Matsuken in 2013, but instead of using recorded voice, this time he used live human voice. Matsuken is the nickname of Matsudaira Ken (in the Japanese naming structure, with the family name first). Matsudaira is a popular actor who often plays samurai roles and is also the singer of a hit series of songs called 'Matsuken Samba', in which Matsudaira sings and dances while wearing samurai costume to music with a pseudo-samba flair. KDDI, a telecommunication company in Japan, had produced a series of popular TV commercials for its product that feature Matsudaira as an android character (with a helmet styled to look like the samurai hairdo) in 2012. Real Android Matsuken was built for an advertisement for another product of KDDI. The android is a talking head of Matsudaira, voiced in real time by comedic impersonator Tomoyuki Yano, with a half torso of a mechanical body that shows its wires and a mechanical skeleton. At the launch event, Matsudaira had a nonsensical conversation with the machine head, moderated by an MC, while Yano comically replied. Although the use of an impersonator's voice turned the android into a remote-controlled robotic puppet, the machine's faultiness, its obviously imperfect relation to the original, became an appealing and hilarious aspect of the presentation.

The successful format of comic conversation between a celebrity and his or her android double/impersonator took on a new twist in the case of Matsukoroid. This celebrity android double, which appears in a seated position, is modelled on popular TV talk-show host Matsuko Deluxe, and it was used in *Matsuko and Matsuko*, a variety TV show appearing on Nippon Television Network Corporation in 2015. Matsuko Deluxe is an outspoken, gay cross-dresser with a very large body. It is visually amusing when the two Matsukos appear on TV side by side, both beautifully attired, and the performance's comic effects are redoubled by the voice of impersonator Hori. Matsukoroid's limited movements (of the face, head, and neck) become part of the robot's 'normal' appearance, its expected persona, in the context of the hilarious and outrageous conversation between the two seated Matsukos. Its deadpan expression makes the more outrageous aspects of their exchange especially funny. The robot's movements synchronise well with the impersonated voice, in the same way that the chanter's voice in Bunraku performance works with Bunraku puppets. It is the combination of the comedic format, the stationary presentation of the two 'stars', and the vital force of Hori's voice that generate the presence that the audience may attribute to Matsukoroid, turning the strangeness of the otherwise large and clumsy remote-controlled android into something that can be accepted socially.

Ishiguro's series of android events at department stores were experiments in ways to present Ishiguro's anthropomorphic machines as able to communicate in a 'real' social situation. Ishiguro's android was used in non-theatre spaces, with staged 'scenarios' that generated an adequate 'narrative' framing (or, Power's fictional mode of presence) for the task at hand. When the cultural frame of celebrity was added into the mix, Ishiguro's android doubles could capitalise on the aura granted by fame, as in the case of Beicho Android. For Real Android Matsuken, the android's persuasiveness was enhanced by the vocalic power of the impersonator, while it made social and media 'sense' enmeshed in a matrix of cultural parody on several levels: a parody of the actor himself, known for his samurai roles; the self-reflexive humour of his fake samba songs, referencing his stage persona, as Matsudaira was costumed in a glamorous version of what he may have worn in samurai epics; and then Matsudaira's performance in the TV ad, singing his samba songs while dressed like a samurai robot. All of these representations, even those that were not present, would be metonymically called into play when either Matsudaira or the android appeared on television. In addition to the factors of celebrity and comic impersonation, Matsukoroid has been successful because it is set within the *mise en scène* of a Japanese TV variety show, which can accommodate anything unusual or extraordinary as long as it is funny (or outrageous). Matsukoroid as the android double of a TV personality functions as a 'character', making appearances on the programme in its own right. Ishiguro's android experiments, particularly those that participated in televised Japanese popular culture, show that his anthropomorphic machines can be successfully assimilated into both fictive and quasi-social narratives, given the right cues, setting, and visual and relational set-up, what I have been referring to as the theatrical *mise en scène*, the totality of staging and performance in public robot displays and situations. As well—a matter of relevance to my discussion in a later chapter on Hatsune Miku, a software character—these android 'personalities' have successfully entered the larger matrix of characters (via their televisual images) in circulation in Japanese popular culture in the same way that manga/anime characters have become part of the popular-culture lexicon of the robot.

Cultural Affordance and Preconceived Outcomes

It is useful to consider some culturally specific notions that predispose the Japanese engagement with humanoids in the situations I have described. I refer to the concept of 'affordance', as psychologist James J. Gibson uses it (1979). It is a coined term to describe how an animal 'reads' information from the surrounding environment in order to determine its possible actions. The term can be used to establish the ways in which human response can be shaped by object design or environmental surroundings. Cognitive scientist and psychologist Donald Norman applies the term in human-machine interaction, stressing the importance of the perceiver's past experiences (1988). The situation of the TV variety show, for example, provides a kind of affordance or, as I will use the concept of affordance in this chapter, a set of recognised, environmental terms and related codes, that condition audience response. As I explained, the TV variety show format is sufficiently flexible in its privileging of the unusual (where there may be a comic payoff) to 'afford', or allow for, the seamless, and successful, incorporation of a talk-show-style 'conversation' involving Matsukoroid.

I would now like to briefly discuss other kinds of cultural affordance in relation to the robot and android performances of Hirata and Ishiguro.

As I discussed earlier in this chapter, emphasising the right timing and the use of onomatopoeia in the Japanese language, Hirata's delivery of dialogue is meticulously calculated to achieve desired effects, in tandem with the other aspects of his theatre 'language'. Hirata's plays deal with issues and concerns familiar to contemporary Japanese, and this topicality also functions as a form of cultural affordance. Hirata mobilises disparate theatrical elements in an economy of known signs regarding the robot that are highly legible, and even attractive, to the Japanese audience, another way of talking about culturally specific affordance. Topics such as '*hikikomori*' and 'the Fukushima disaster' are intended to attract the Japanese audience's attention, and they are meant to be poignant, eliciting emotional response and a sense of urgency.

It is easy for Western critics to miss the importance of these pointedly Japanese references in Hirata's plays with robots and androids. Nestruck, for example, criticises Hirata's dramaturgy in *I, Worker* and *Sayonara* as 'hopelessly old-fashioned in its human/machine segregation' because, in his view, these productions 'typecast' robots: actual robots are given the roles of robots, unlike 'cyborg performance art', which may deploy robots very differently; as well, he feels 'our love-hate relationship with our robotic servants' is inadequately explored (2013). His comments are understandable from the Western perspective that I discussed earlier, in which the robot is seen highly ambivalently. I have also remarked upon the criticisms that Hirata's plays have received concerning gender stereotyping. However, the fact that the roles of carer robots in Hirata's plays are performed by 'female' robots can be viewed as reflecting the norm in the nursing home and aged care industries in Japan, and in the home.[10] Western critics are, of course, also missing the governmental directives that I raised earlier in this chapter and in the previous chapter, which predispose robot projects to focus on harmonious relations between robots and humans, a prerequisite for robot integration into homes and institutions.

In summary, the robot and android theatre works by Hirata and Ishiguro, as well as Ishiguro's android experiments, operate through systems that seek to integrate robots into social (or media) situations, and they are for the most part quite successful at doing so because they are adept at structuring contextual frames and at deploying Japanese cultural cues, particularly those that have affective weight. What might seem 'mawkish' for the Western critic is more likely to be pointed and meaningful for the

Japanese audience. It therefore makes sense that they might wish to mini-mise or hide the non-human qualities of these machines. Artist Kenji Yanobe, on the other hand, is not afraid of robots as highly ambiguous emblems of misguided technological fantasy. In the final section of this chapter, I examine works suggesting a position on robot culture that is antithetical to that of Hirata and Ishiguro. I examine Yanobe's robotic installation works *Viva Riva Project: New Deme* (2002) and *Giant Torayan* (2005). These works use robots in humorous ways, as for Ishiguro's Matsukoroid, but they reject assimilative tendencies and party-line politics.

ROBOTS AND DISILLUSIONMENT

Kenji Yanobe is a prominent artist who became known in the 1990s for his performances, sculptures, and installations that reflected a disenchanted, even apocalyptic, view of technology. In his works, Yanobe responds with biting humour to technological folly and technology-related disasters, such as the Chernobyl nuclear accident. Yanobe's works are discussed in terms of the sense of crisis felt in Japan during the 1990s due to the economic recession that followed the 'bubble' economy of the 1980s and disastrous events such as the Great Hanshin earthquake and the Tokyo subway sarin attack, both in 1995 (Borggreen 2006, 128).

One of Yanobe's earlier works, *Foot Soldier (Godzilla)* (1991), is meant to be a bipedal transportation machine, consisting of an open cockpit in which one might sit, above Godzilla's lower torso and legs, which are deep blue and motorised, with two supporting metal poles with a wheel at each end. While Yanobe claims that this vehicle can roam like the eponymous monster, it is meant to be ironic, making the viewer 'laugh and feel a little nostalgic' (Nukada and Murakami 2005, 65). *Yellow Suit* (1991), on the other hand, consists of a massive nuclear protection suit made of iron and lead that is able to clothe both himself and his dog, a despairing comment on the ridiculous and extreme steps one might need to take to protect oneself and one's loved ones from a situation that should never have happened, that was itself ridiculous. The work was made in response to a nuclear accident in 1991 at Mihama nuclear plant, which is 80 km northeast of Kyoto. In *Atom Suit Project* (1997), wearing a yellow protection suit equipped with Geiger counters and a yellow diving helmet featuring two Astro-Boy-like pointy cones attached to it, Yanobe photographed himself at abandoned sites such as an amusement park and a kindergarten near the

Chernobyl nuclear plant, which had been crowded with people before the disaster. It isn't such a bright future after all, Yanobe's work indicates.

The following year, Yanobe took *Atom Suit Project* to the former site of the 1970 exposition in Osaka as a reminder of 'the leftovers of that imagined future' (Borggreen 2006, 119), a pointed comment on the naiveté of those imaginings. The exposition site held personal significance for Yanobe, as he had lived nearby as a child during the early 1970s and was quite familiar with the abandoned site that was once full of spectacular buildings and grand pavilions. This childhood experience of seeing the discarded site, including Isozaki's giant robot Deme, left a long-lasting impression (Sawaragi 2002, 28). The expo had served to create 'a sense of expectation and mission regarding the land of the future' that 'had sadly met its end and become a vacant lot' (Oba 2013b, 67, original English). The sense of ruined dreams and the demise of great hopes that Yanobe sees in the vacant site echoes the sentiments felt in the post-Expo period in Japan, the late 1980s and 1990s. Art critic Noi Sawaragi describes how quickly expectations turned into disillusionment due to a confluence of economic, social, and environmental crises:

> As *Expo '70* was underway, a radical New Left group hijacked a domestic aircraft and the novelist Yukio Mishima staged his suicide by traditional disembowelment. In the next few years, a series of terrorist bombings hit downtown Tokyo, President Nixon's suspension of the gold standard and introduction of fluctuating currency exchange rates provoked the 'dollar shock,' and the international oil crisis precipitated the 'oil shock,' which in turn caused spiralling inflation. These events spurred a national doubt that the promised bright future would ever arrive. These years also saw environment crises plague the whole nation, with city children regularly advised against outdoor exercise because of air pollution. A new kind of pessimism was pervasive, even amongst children. (2005, 192, original English)

Robots are often used as 'guides' to these ruined futures in Yanobe's work. For example, in 'Expose 2002 – Far Beyond Dreams of the Future', an exhibition held in Osaka in 2002 that intended to re-evaluate the aims of Expo '70 (hence the similarity of the titles and the play on the word 'Expo', with the implication of critical examination for the 2002 show), Yanobe combined themes explored in the previous projects in robotic works titled *Viva Riva Project: Standa* (2001) and *Viva Riva Project: New Deme* (2002) (Yanobe 2002). These two installation works prominently featured large-scale sculptures that faced each other in the exhibition space, repeating a disturbing 'conversation' again and again.

Viva Riva Project: Standa was an installation that consisted of a three-metre-tall robotic baby doll, a blown-up, metallic version of a doll that Yanobe picked up at an abandoned preschool near Chernobyl. The large doll had an aluminium and brass surface. On the wall behind the giant, metallic baby doll was the shape of a sun, with sawtoothed rays and a smiling face. The doll, initially bent forward, stood up whenever it 'detected' radiation 20 times, and the sun-image would then glow, radiating light as if in celebration. *Viva Riva Project: New Deme* was an homage to Isozaki's doomed performing robot Deme at the 1970 Expo, which I discussed in Chap. 3. While Yanobe's three-metre-high New Deme was equipped with the same distinctive pop-eyed features of the original Deme, it had a round body of a brownish rusty colour, with two arms, two legs, and a tail, looking like a kind of insectile submarine with two baby-like, short legs. With its exposed metal frame and rivets, New Deme had a definite steam-punk look. When the Standa robot stood up, New Deme prostrated itself on the floor, submerging its 'face' in a tank of water. The resonance of past with present was like a 'simulated experience of time travel' (Oba 2013b, 67). It was as if Standa and New Deme came back from the ruined site of the past with new robotic bodies that revealed the misplaced optimism of their originals, and entered into a perpetual dialogue about blindness to disaster and momentary repentance.

The idea of the robot as a fantasised medium beyond time and form, and as a reminder or ghost from the past that perversely reveals something of Japan's reluctance in the present to let go of its once-gleaming dreams for the future, continued to be developed in Yanobe's robotic works. In 2004, Yanobe was asked to produce an art project during the 2005 Aichi Exposition by one of its sponsors, The Chūnichi Shimbun, a newspaper company based in the Aichi region. Yanobe proposed to create a robot mammoth as his response to an exhibition at the expo in which a frozen head of a mammoth was to be exhibited as one of its major attractions. Yanobe's idea of the robot mammoth was a four-legged monster, 20 metres long and weighing 20 tons, with a body made of metal scraps, which was powered by a diesel engine and emitted thick, black smoke as it walked. After its launch at the exposition site, Yanobe planned to take it to the city of Nagoya by helicopter and let it walk from the city to the port. The robot mammoth was then to be shipped to Siberia and buried for preservation. While this idea was rejected by the sponsoring company, Yanobe realised part of his original plan in *Mammoth Pavilion* (2005), in which a metal replica of a baby mammoth was presented in ice, and in *Rocking Mammoth* (2005), a large (about three metres in height, to its back) mammoth-shaped rocking chair made of parts taken from his car, a 'Toyota High Ace', as he put it (Yanobe 2005). The ironic use of the term 'pavilion' and the reference to Toyota in relation to the Aichi Exposition, as Toyota was based in the Aichi prefecture, are obvious.

Another robot of Yanobe's where the irony is hard to miss is his *Giant Torayan* (2005). Torayan is Yanobe's name for a fictional character modelled on a ventriloquist's dummy owned by his father. The character's appearance, like that of the dummy, is something between a child and an adult—a child-sized, round body with a baby face that has a moustache and a middle-aged man's comb-over with a few hairs (Oba 2013a). Yanobe referenced this character in an installation work in 2004, titled *Cinema in the Woods*, in which Yanobe screened a film of his father explaining how to survive in wartime to the ventriloquist's dummy on which Torayan is based; the dummy was dressed in the protective suit from Yanobe's *Atom Suit Project*. *Giant Torayan* is one of the variations of the *Torayan* series, which was conceived as an alternative to the rejected mammoth project series, though the idea of *Giant Torayan* was also rejected by the newspaper company. Yanobe nevertheless actualised it in an exhibition in the following year, 2005, along with *Rocking Mammoth*. Giant Torayan is a 7.2-metre metal giant, made of aluminium, steel, and brass. It talks with a taped voice, and opens and closes its mouth and eyes, twists its head, and dances by moving its arms and twisting its body. The robotic machine also spews fire from its mouth on command.

Giant Torayan is a robotic figure that is at the same time familiar and friendly, as is a child's doll, but is also alien and frightening, as a doll may be when it is found abandoned next to a school in Chernobyl or when it is a father's avatar for his wartime survival. Japanese studies scholar Suzan Napier discusses Godzilla in terms of the Japanese audience's fear and fascination with destructive monsters as a result of the devastation and trauma of World War II (1993, 349).[11] Yanobe's Giant Torayan is such an ambivalent figure, not unlike the ambivalent, technological powers embodied in the manga figure of *Iron Man No. 28*. But Giant Torayan is not a menacing character, as in the Western imagination, such as the robot of *Terminator*. Yanobe's robot is a trickster, darkly humorous and mischievous, a blown-up version of a doll and one that is full of surprises, firing flames from its mouth. It embodies technology as the return of the repressed, banished by the sanitised appearance and performance of next-generation robots. The power and violence of robot technology leaps out as fire. The audience would recognise both the attractions of the giant baby robot as a robot and note its jaundiced view of humanoid robotics—silly, pointless, dangerous, overblown.

The modality of installation allows for more flexible viewing perspectives than that of theatre. Gallery viewers have more freedom to enjoy the jester character of Giant Torayan without being directed to take a particular narrative perspective, as in Hirata and Ishiguro's carefully controlled theatre—and as in ideologically driven government and corporate statements as they are communicated in their theatre and in demonstration exhibitions and shows.

Yanobe is critical of the progressivist view of technology associated with Japan's post-war development. In Yanobe's work, the figure of the robot humorously gestures to the nation's helpless implication in the vagaries of a post-war ideology of prosperity through science and technology. Sociologist Munesuke Mita uses the terms 'ideal', 'dream', and 'fictionality' to describe particular periods of the post-war era in Japan. The first term refers to the reconstruction period from the end of the war to the end of the 1950s; the second, to the high-growth period, also a time of radical-left politics and protest, from 1960 to the mid-1970s; and the third term refers to the period of the bubble economy and an expanding consumerism from the late 1970s to the end of the 1980s. Mita posits these terms as the antithesis of the actual 'reality' of these three periods as they were experienced by the Japanese people (quoted in Uno 2011a, 416). In his locating of the future in the past and the past in the present, Yanobe's robotic works reflect a mixing of 'ideal', 'dream', and 'fictionality' and their related disenchantments.

Notes

1. According to roboticist Jun'ichi Takeno, a half second is the maximum time one can allow before humans feel a delay in an action (2011, 17).
2. Hirata's view echoes that of Christian Denisart, a Swiss theatre director and playwright, who produced *Robots* in 2009, a theatre production with interactive robots. Denisart acknowledges that his production faced technical limitations where the robots were concerned (Saltz 2015, 113).

3. Hirata's views on the actor and acting resonate with those of Heinrich Kleist, a German dramatist of the late eighteenth century, and Edward Gordon Craig, an English modernist theatre director and theorist. They suggest replacing an actor with a puppet or a marionette, an approach that can overcome human limitations. A similar view has also been put forward by Japanese dramatist of *ningyō jōruri* and Kabuki from the Edo period, Monzaemon Chikamatsu.

4. Hirata's view echoes that of research psychologist Elly Konijn, who empirically highlights the importance of 'task-emotion' in stage acting, disproving the long-held view of a unity of emotion between actor and character (2000).

5. The use of remote-controlled robots for research on human-robot interaction is known as the 'Wizard of Oz (WoZ)' technique. It raises methodological concerns among HRI (human-robot interaction) researchers because 'it is not really human-robot interaction so much as human-human interaction via a robot' (Riek 2012, 119).

6. There have been variations of this production: for example, the main female character has been played by a Japanese actor; English has been used as the main language, with poems recited in French, in German, and in Japanese; the use of a recorded voice has been used for Geminoid F; and the production has used projections of translated texts.

7. As Hirata's plays portrayed the robots as domestic companions, and thus domestic tools or facilitators, it is possible to imagine that the Japanese audience of the Ishiguro-Hirata collaborative works project such views onto the robots on stage.

8. Hirata also refers to the Japanese tradition of animism in terms of a blurring of the boundary between humans and robots (quoted in Doherty 2013).

9. The Turing Test was introduced by British pioneering computer scientist and mathematician Alan Turing in 1950 to consider machine intelligence and self-identity in relation to those of a human.

10. Online debates concerning a cover image of *The Journal of the Japanese Society for Artificial Intelligence* published in 2014 suggest that the idea that the 'gender' of the carer robot should be female is still common. This journal was criticised when it used an illustration of a female android as a housekeeper on its cover (Yamada 2014). It

should be remarked, however, that it is not only in Japan that the gender preference for caregivers is female. Psychologist Julie Carpenter and colleagues have conducted an experiment on gender representation relating to robots for domestic use, using Ishiguro's female robot Repliee. The participants, ten female students and nine male students at the University of Washington in the USA, disclosed 'preferences for a female robot for in-home use' (2009, 263). Glenda Shaw-Garlock, a researcher on society and technology, points out that designers of social robots internationally like to use gender stereotyping as an easier and more predictable option to establish communication between a user and his or her machine (2014, 313). (I will come back to this issue of gender-based roles in aged care in Chap. 8.)

11. *Shin Gojira* (English title, *Shin Godzilla* or *Godzilla Resurgence*), co-directed by Hideaki Anno and Shinji Higuchi and screened in 2016, depicts the government's inept response at a time of crisis as an allegory of the Fukushima nuclear disaster in 2011.

References

Asada, M. (2010). *Robotto to iu sisō* [A philosophy called the robot]. Tokyo: NHK Shuppan.

Borggreen, G. (2006). Ruins of the future: Yanobe Kenji Revisits Expo '70. *Performance Paradigm: Journal of Performance and Contemporary Culture, 2,* 119–131.

Borggreen, G. (2014). 'Robots cannot lie': Performative parasites of robot-human theatre. In *Sociable robots and the future of social relations: Proceedings of Robo-Philosophy 2014* (pp. 157–163). Amsterdam: IOS Press.

Carpenter, J., Davis, J. M., Erwin-Stewart, N., Lee, T. R., Bransford, J. D., & Vye, N. (2009). Gender representation and humanoid robots designed for domestic use. *International Journal of Social Robotics, 1*(3), 261–265. doi:10.1007/s12369-009-0016-4.

Connor, S. (2000). *Dumbstruck: A cultural history of ventriloquism.* New York/Oxford: Oxford University Press.

D'Cruz, G. (2014). 6 things I know about Geminoid F, or what I think about when I think about android theatre. *Australasian Drama Studies, 65,* 272–288.

Doherty, M. (2013, March 2). With I, Worker, Canadian Stage takes on the inevitable robopocalypse. *National Post.* http://news.nationalpost.com/arts/on-stage/with-i-worker-canadian-stage-takes-on-the-inevitable-robopocalypse. Accessed on 10-10-2014.

Fischer-Lichte, E. (2014). *The Routledge introduction to theatre and performance studies.* M. Arjomand & R. Mosse (Eds.) (M. Arjomand, Trans.). London/New York: Routledge.

Gibson, J. J. (1979). *The ecological approach to visual perception.* Boston: Houghton Mifflin.

Hibino, K. (2012). Oscillating between Fakery and Authenticity: Hirata Oriza's Android Theatre. *Comparative Theatre Review, 11*(1), 30–42. doi:10.7141/ctr. 11.30.

Hirata, O. (1995). *Gendai kōgo engeki no tameni* [For contemporary colloquial theatre]. Tokyo: Banseisha.

Hirata, O. (1998). *Engeki nyūmon* [Introduction to theatre]. Tokyo: Kōdansha.

Hirata, O. (2004). *Engi to enshutsu* [Acting and directing]. Tokyo: Kōdansha.

Hirata, O. (2012). About our Robot/Android Theatre (K. Hibino, Trans.). *Comparative Theatre Review, 11*(1): 29. doi:10.7141/ctr.11.29.

Hirata, O., Ishiguro, H., & Kinsui, S. (2010). Robotto ga engeki? Robotto to engeki!? [A robot in the theatre? Performing with a robot!?]. In Osaka daigaku komyunikēshon dezain sentā (Ed.), *Robotto Engeki* (pp. 14–33). Osaka: Osaka Daigaku Shuppankai.

Hotta, J. (2008). *Hito to robotto no himitu* [The secret concerning robots and humans]. Tokyo: Kōbunsha.

Ishiguro, H. (2007). *Andoroido saiensu: Ningen wo sirutame no robotto kenkyū* [Android science: A study to learn what the human is]. Tokyo: Mainichi Komyunikēshonzu.

Ishiguro, H. (2009). *Robottoto wa nanika: Hito no kokoro wo utsusu kagami* [What is the robot?: A mirror reflecting the human soul]. Tokyo: Kōdansha Gendaishinsho.

Ishiguro, H. (2011). *Dōsureba hito wo tsukureruka: Andoroido ni natta watashi* [How can a human be made: I became an android]. Tokyo: Shinchōsha.

Ishiguro, H. (2012). *Hito to geijutsu to andoroido: Watashi wa naze robotto o tsukurunoka* [The human being, the arts, and the android: Why do I make androids?]. Tokyo: Nihonhyōronsha.

Ishiguro, H. (2013). *Kusobukuro no uchi to soto* [Inside and outside the human body]. Tokyo: Asahi Shimbun Shuppan.

Ishiguro, H., & Ikeya, R. (2010). *Robotto wa namida wo nagasuka* [Does a robot Shed Tears]. Tokyo: PHP Kenkyūjo.

Ishiguro, H., & Oriza, H. (2015). Ishiguro Hiroshi X Hirata Oriza: Aondoroido wa ningen no yume wo miruka [Can an android have dreams like humans do]. In Bungē Bessatsu (Ed.), *Hirata Oriza: Sōtokushū, sizukana kakumei no kishu* [Special issue, the leader of the quiet revolution] (pp. 118–125). Tokyo: Kawade shobō.

Konijn, E. (2000). *Acting emotions: Shaping emotions on stage* (B. Leach, Trans.). Amsterdam: Amsterdam University Press.

Kuroki, K. (2010). Robotto engeki no kaihatsu [The development of robot theatre]. In Osaka daigaku komyunikēshon dezain sentā (Ed.), *Robotto engeki* [Robot theatre] (pp. 60–65). Osaka: Osaka Daigaku Shuppankai.

Matsumoto, K. (2015). *Hirata Oriza: Sizukana engeki to iu hōhō* [The methodology of quiet theatre]. Tokyo: Sairyūsha.

Mizu'ushi, K. (2011, October 5). Hirata Oriza X Ishiguro kenkyūshitsu, Andoroido engeki, Sayonara [Hirata Oriza X Ishiguro Laboratory, Andoroido Theatre, Sayonara]. Wonderland. http://www.wonderlands.jp/archives/18906/.

Nagata, T. (2005). *Robotto wa ningen ni nareruka* [Can a robot be a human being]. Tokyo: PHP Kenkūsho.

Nakanishi, O. (2014). Kurosu rebyū: Hirata Oriza + Ishiguro kenyūshitsu, andoroido engeki, Sayonara [Cross-review: Hirata Oriza + Ishiguro Laboratory, Android Theatre, Sayonara]. *Kokusai Engekika Kyōkai Engeki Hyōronshi, ACT* 21. http://act-kansai.net/?p=73. Accessed on 10-02-2015.

Napier, S. J. (1993). Panic sites: The Japanese imagination of disaster from Godzilla to Akira. *Journal of Japanese Studies, 19*(2), 327–351. doi:10.2307/132643.

National Institute of Informatics. (2011). Creating robot theater for building a more preferable robot. *KAKEN: Database of Grants-in-Aid for Scientific Research.* https://kaken.nii.ac.jp/d/p/23240027.ja.html. Accessed on 07-15-2014.

Nestruck, J. K. (2013, February 27). Sayonara, I, worker: These plays are a little too robotic. *The Globe and Mail.* http://www.theglobeandmail.com/arts/theatre-and-performance/theatre-reviews/sayonara-i-worker-these-plays-are-a-little-too-robotic/article9123716/. Accessed on 12-10-2014.

Norman, D. A. (1988). *The psychology of everyday things.* New York: Basic Books.

Nukada, H., & Murakami, T. (2005). Little boy (plates and entries). In T. Murakami (Ed.), *Little boy: The arts of Japan's exploding subculture.* New York/New Haven/London: Japan Society and Yale University Press.

Oba, M. (2013a). Cinema in the wood. In K. Yanobe (Ed.), *Yanobe Kenji: 1969–2005* (p. 111). Kyoto: Seigensha.

Oba, M. (2013b). The ruins of the future. In K. Yanobe (Ed.), *Yanobe Kenji: 1969–2005* (p. 67). Kyoto: Seigensha.

Okuno, T. (2002). *Ningen dōbutsu kikai: Tekuno animizumu* [Human, animal, machine: Techno animism]. Tokyo: Kadokawa shoten.

Omori, S. (1981). *Nagare to yodomi: Tetsugaku danshō* [Flow and stagnation: Philosophical fragments]. Tokyo: Sangyō Tosho.

Parker-Starbuck, J. (2015). Cyborg. Returns: Always-already subject technologies. In S. Bay-Cheng, J. Parker-Starbuck, & D. Z. Saltz (Eds.), *Performance and media: Taxonomies for a changing field* (pp. 65–92). Ann Arbor: University of Michigan Press.

Riek, L. D. (2012). Wizard of Oz studies in HRI: A systematic review and new reporting guidelines. *Journal of Human-Robot Interaction, 1*(1), 119–136. doi:10.5898/JHRI.1.1.Riek.

Saltz, D. Z. (2015). Sharing the stage with media: A taxonomy of performer-media interactions. In S. Bay-Cheng, J. Parker-Starbuck, & D. Z. Saltz (Eds.), *Performance and media: Taxonomies for a changing field* (pp. 93–125). Ann Arbor: University of Michigan Press.

Sasaki, A. (2011). *Sokkyō no kaitai/kaitai: Ensō to engeki no aporia* [Dismantling and conceiving improvisation: Aporia in playing musical instruments and in theatre]. Tokyo: Seidosha.

Sawaragi, N. (2002). *Expose 2002: Far beyond Dreams of the Future, Kenji Yanobe X Arata Isozaki.*

Sawaragi, N. (2005). On the battlefield of 'Superflat': Subculture and art in postwar Japan. In M. Takashi (Ed.), *Little boy: The arts of Japan's exploding subculture* (L. Hoaglund, Trans.) (pp. 187–207). New York/New Haven/London: Japan Society and Yale University Press.

Searle, J. R. (1980). Minds, brains, and programs. *Behavioral and Brain Sciences, 3* (03), 417–424. doi:10.1017/S0140525X00005756.

Sena, H. (2008). *Sena Hideaki robottogaku ronshū* [Hideaki Sena robot study essay collection]. Tokyo: Keisō Shobō.

Shaw-Garlock, Glenda. (2014). Gendered by design: Gender codes in social robotics. In *Sociable robots and the future of social relations: Proceedings of robo-philosophy 2014* (pp. 309–317). Amsterdam: IOS Press.

Sonoyama, T. (2007). *Robotto dezain gairon* [Introduction to robot design]. Tokyo: Mainichi Komyunikēshonzu.

Takeno, J. (2011). *Kokoro wo motsu robotto: Hagane no shikō ga kagami no nakano jibunni kizuku* [The robot with soul: A realisation of the self in the mirror by a metallic thinking entity]. Tokyo: Nikkan Kōgyōsha.

Uno, T. (2011a). *Ritoru pīpuru no jidai* [The age of little people]. Tokyo: Gentōsha.

Varney, D., Eckersall, P., Hudson, C., & Hatley, B. (2013). *Theatre and performance in the Asia-Pacific: Regional modernities in the global era.* Houndmills/Basingstoke/Hampshire/New York: Palgrave Macmillan.

Yamada, H. (2014, January 8). Jinkō chino gakkai no ayamari [The problem caused by the Japanese Society for Artificial Intelligence]. *The Huffington Post Japan.* http://www.huffingtonpost.jp/hajime-yamada/post_6588_b_4560115.html. Accessed on 10-10-2015.

Yanobe, K. (2002). Ano Deme ga kaettekita [Deme Has Returned]. In Kirin Puraza Osaka (Ed.), *Expose 2002* (pp. 26–27). Osaka: Kirinbīru KPO Kirin Puraza Osaka.

Yanobe, K. (2005). Rocking Mammoth. *Kenji Yanobe Archive Project.* http://www.yanobe.com/artworks/rockingmammoth.html. Accessed on 05-01-2015.

CHAPTER 5

Robots, Space, and Place

This chapter examines Japanese entertainment presentations featuring 'fighting' humanoid robots. The popular image of the fighting robot, typically a standing warrior figure wearing a helmet and body armour and holding a weapon, has become naturalised among the post-war generations of Japanese through anime seen on television and in manga. As I discussed in the previous chapters, some robot characters, such as Astro Boy, in essence an early fighting robot, have achieved iconic status. I will discuss how the meaning of the fighting robot is created through the spectator's active interpretation, given the prompts of their particular external contexts. Unlike Kenji Yanobe's art installations, which overtly express ambivalence regarding technology through the figure of the robot, case studies look at popular forms of the genre that are intended to entertain an audience. The works I examine include a statue of Iron Man No. 28 (a robot animation character); a daily multimedia performance involving a Gundam (a robot animation character) statue in front of a shopping mall complex; a demonstration show of Kuratas, an exoskeletal humanoid robot at a large hobbyists' festival; and a techno-fantasy cabaret show for foreign tourists at a venue called Robot Restaurant in Shinjuku, a known entertainment and red-light district in Tokyo. I examine how these popularised productions develop narratives of the 'real' and fictive Japan in relation to technology, expressing what I will explain as a self-reflexive Orientalism.

These popular fighting robot performances are designed for varied audiences, relative to their specific locations: the general public, fans of giant-robot manga/anime, and foreign tourists, respectively. To examine the

© The Author(s) 2017
Y. Sone, *Japanese Robot Culture*,
DOI 10.1057/978-1-137-52527-7_5

ways in which location matters in these performances, I consider philosopher Michel de Certeau's notion of place. For de Certeau, a place is an 'ordered' and regulating system for stability, and '[t]he law of the "proper" rules in the place' (1984, 117). In contrast, a space is regarded as '*a practised place*' that exists through multiple expressions of spatial activities and is indifferent to 'the univocity or stability of a "proper"' (de Certeau 1984, 117, original emphasis). In this chapter, I am primarily concerned with articulations of location that have already been determined by external factors.

The statues of Iron Man No. 28 and Gundam as, essentially, public art are designed to affect their locales in the manner of 'place', demanding the viewers' recognition and appreciation of their roles in a regulated and sanctioned public exchange. The material conditions of the robot statues and the locations of their installations become meaningful in establishing and maintaining place. The sheer scale and material presence of the Iron Man monument and the Gundam statue give rise to a 'wow-effect', and, at the same time, their literalness limits imaginative scope for their viewers, so that commercial narratives are not derailed. The uses of these images of the robots they portray are dramatically enacted in these particular sites for commercial gain, adding value to their locations and helping to define place. On the other hand, the contexts of the debut demonstration performance of Kuratas and of the Robot Restaurant are already determined by the meanings associated with their locations, as known 'places', in Tokyo. The context and venue of Kuratas's show, Tokyo Wonder Festival and Makuhari Messe, are well known within fan culture. Shinjuku, where Robot Restaurant is located, is a recognised entertainment district in Tokyo.

GIANT ROBOT STATUES

In 2009, in order to attract visitors to Kobe after the reconstruction that followed the disaster of the 1995 Great Hanshin Earthquake, a 15-metre high, 50-ton robot statue of the main character of the eponymous *Iron Man No. 28* (a 1956 manga written by Kobe-born Mitsuteru Yokoyama) was erected in western Kobe. The Iron Man statue was intended as a permanent monument in the Wakamatsu Park near Nagata station for the municipality's revitalisation project for devastated local industries. Unlike a typical statue of a human figure, the pose of the Iron Man monument is not designed to create a relatively static image but to be dynamic and uplifting. Its pose is that of a boxer punching an opponent. The statue stands with its two legs apart, with the left leg bent and the right straight, and its right arm straight out at 90 degrees with a closed fist, while the left fist seems ready for the next blow. The Iron Man statue is facing the shopping district that was severely damaged by the quake, as if keeping a very close watch, ready to defeat the forces of nature. One might assume that the producers of the statue hoped that it would be regarded as a kind of guardian deity, like the statues of Buddhist demon gods at temples, as well as becoming a popular tourist attraction.

As a public monument for local economical recovery, the Iron Man statue conveys a clear message of goodwill. It also reflects a widespread view in developed nations that '[p]ublic art is good' because it '"enhances the quality of life" or "humanizes the urban environment"', as ironically described by philosophers Douglas Stalker and Clark Glymour (1982, 4). According to cultural theorist Malcolm Miles, public art and the monument 'define and make visible the values of the public realm, and do so in a way which is far from neutral, never simply decorative' (Miles 1997, 61). The use of a contemporary hero for this statue, a figure from a popular manga/anime, represents an updating of the figure of the guardian deity in terms that all could appreciate, but it also reflects the established techno-utopian narrative since Astro Boy of the robot as saviour of Japan. The considerable size of the animation figure in a real public space is captivating, demanding that its literal presence is felt as an island of ideological compulsion, a 'place', following de Certeau, in the midst of what might possibly be, for the most part, a space. In the midst of a public park, to be used by residents as they wish, the statue embodies a kind of directive to look to the future after the disaster, and the future is about the development of robots (and the development and export of Japanese popular culture, a form of soft power of greater relevance in the long period following the end of the economic boom).

The statue of Gundam is another case of a carefully engineered spectacle facilitated by commercial interests. It is a much more complex case of a performing object that makes meaning within a 'transmedia' franchise (Jenkins 2006, 95).[1] The franchise covers a large number of popular anime series as well as films, and both have been ongoing on Japanese television since the first series, called *Mobile Suit Gundam*, premiered in 1979. It is one of the most popular robot anime. Its basic plotline involves gigantic humanoids, usually piloted by teenagers, fighting enemies in space. The *Gundam* anime became more popular following the introduction of plastic models of Gundam machines by Bandai, a large Japanese toy company, in the 1980s. In 2009, a statue of a Gundam robot was erected for two months in the summer on Odaiba Island, a major commercial, residential, and leisure area in Tokyo Bay. It was built in 1:1 real-size scale to the measurements the robot is supposed to have according to the anime: 18 metres high, like the Iron Man statue. Unlike the statue of Iron Man. which remained motionless, this robot figure was able to move its head, expel a steam mist from parts of its body, and illuminate its eyes and other parts of its body at night. The statue was built to mark the 30th anniversary of the television broadcast of *Mobile Suit Gundam* in 1979 as well as to support Tokyo's bid for the 2016 Summer Olympics (under the rubric 'Green Tokyo Gundam Project').[2] It was reported that more than four million people visited it in a period of 52 days (*IT Media News* 2009). In the following year, 2010, the statue was re-erected in the city of Shizuoka, where Bandai's model factory is located, to celebrate the thirtieth anniversary of the sale of Gundam plastic models as part of the Shizuoka Model Show, a major annual plastic model exhibition. The Gundam figure—in disassembled parts so that fans could get close enough to touch them—returned to Odaiba in 2011 for the charity fair held for the Great Eastern Japan Earthquake disaster. It was reinstalled in 2012 in front of the large shopping mall Diver City Tokyo, which includes Gundam Front Tokyo, an entertainment complex. The Gundam statue, in conjunction with Gundam Front Tokyo, distinctively constitute Diver City Tokyo's 'brandscape'. Architect and scholar Anna Klingmann explains brandscapes as 'constitut [ing] the physical manifestations of synthetically conceived identities transposed onto synthetically conceived places, demarcating culturally independent sites where corporate value systems materialize into physical territories' (2007, 83). Here, the identity of Bandai is emblematised by the giant robot, and Diver City Tokyo is designated by this branding as the home of a central attraction, Gundam Front Tokyo.

The Gundam statue 'performs' for Gundam Front Tokyo as a contemporary version of the *maneki neko* (beckoning cat), a common talisman at the entrance to Japanese shops in the form of a cat figurine. The site is effectively a theme park for the Gundam merchandising empire owned by Bandai, which produces manga, animation, video games, and novelty tie-ins, such as plastic models and toys. Its main attractions include a 360-degree panoramic movie theatre, a 1:1 scale model of the upper half of another Gundam robot, museum exhibits of various artworks related to the Gundam series, and the display of hundreds of Gundam plastic models. The Gundam Front allows fans to engage with the world of Gundam in a physical, material, and eventful way. In particular, the massive Gundam figure, which weighs 35 tons, has an overwhelming material presence and has become the icon for the theme park as well as for Diver City Tokyo. A construction of a large-scale shopping mall is highly political. Just like the Iron Man statue's representing of an accepted narrative of Japan's resilience and ultimate success, the Gundam statue also represents investments, fronting an institutionalised commercialism constituted by the vested interests of state (prefecture) and local government and corporate bodies 'in dialogue with the consumer' (Goodlander 2015, 119). Cultural theorists Scott Lash and Celia Lury argue that in advanced economies, cultural products are produced and received across media platforms and modalities, facilitating an interaction between 'media-things' and 'thing-media', where '[i]mage has become matter and matter has become image' (2007, 9). The Gundam statue is such an example of image and materiality transfused for the good of multiple commercial interests.

The daily multimedia 'enactment' of a scene from a *Gundam* animation at night integrates the statue with a 'sound and light' display: it is illuminated with LED light projections, and framed as the main 'performer' of the show. The Gundam statue is able to move its head along with sound effects and to give off puffs of white steam at appropriate moments in the storyline. At the end of the show, before the Gundam machine in the anime is 'launched' for battle, the human characters of the animation 'appear' (projected images displayed in a video monitor) in the opened hatch in the centre of the Gundam statue's chest, as if they were actually inside the suit. While the statue is a puppet, a prop in a multimedia performance presentation, on another level, it 'performs' in a way that enhances its profile and therefore its commercial utility. Its presence is transformative for fans. The performance enables the interaction of fictional *Gundam* manga/anime with the spectacular 'realness' of the larger-than-life-size figure, not in a theatre or an exhibition space but in an open, quotidian locale outside the mall, a quasi-space shared by tourists. This location is corporately designated (Bandai and Diver City Tokyo are enabling this performance, for specific purposes) and at the same time it is public, an ambiguous situation in which the real Gundam figure (super-real, due to its scale and solidity) is fictively framed, a tangible extension of fantasy.

The notion of '*riaru*' (real) was a key concept for the creation of *Mobile Suit Gundam* (Tane 2010, 15). '*Riaru*' means in this context an emphasis upon details and a concern with the moral and social ambiguities of the 'real world', rather than a clear distinction between 'good' and 'evil'. Unlike previous anime that were regarded as TV manga, with the view that they were for elementary school children, *Gundam* was designed to appeal to adolescents and young adults with themes for a mature-age audience. For example, the background setting for its plot is the overpopulation of the earth, which led to the building of space colonies, and that one of the colonies, the Principality of Zeon, launched a war of independence from the Earth Federation. Unlike the robot anime before it, there is no clear division in this narrative between right and wrong in this conflict. To make the story seem closer to the real world, the mobile suits are positioned as replaceable weaponry, just like tanks or fighter jets, unlike Iron Man No. 28, which is a unique, irreplaceable, and invincible 'super robot'. *Gundam* also uses references to physics and advanced scientific terms to suggest its setting as more realistic. The human pilots fight for survival by manoeuvring their mobile suits skilfully. *Gundam* depicts the absurdity of an endless conflict with 'brutal representations of war and fighting' (Condry 2013, 126). While the Japanese ideology for the robotic technology shown in *Iron Man No. 28* is very black-and-white, *Gundam* provides long and complex narratives that reveal the compromises and difficult decisions of actual wars.

Through its epic story, *Gundam* retells the story of World War II. It is a common view among the *Gundam* fans that Zeon is modelled on Nazi Germany because of its idea of a superior 'spacenoid race' (that is, a 'race' of people who developed differently by living in space colonies) and its references to Hitler, according to Kiyoshi Tane (2010, 17–9). While the employment of new technology (Zaku, the first mobile suit) leads to Zeon's earlier success in the war of secession, the Earth Federation, like the Allied forces of World War II, overcomes Zeon's resistance with its superior manufacturing capabilities and greater numbers. Gundam is the name of the test model for the Federation's mobile suit. Despite its numerous Nazi references, as writers such as Tane (2010) and Masayuki Endo (2002) indicate, it is possible to see that the *Gundam* story allegorically reflects Japan's defeat by the USA, due to the latter's overwhelming mass-production capacity. These writers also suggest that *Gundam* reflects Japan's post-war obsession with advance technologies and mass production through industrial robotics. Given both of these interpretations, it seems clear that *Gundam* embodies a deeply felt Japanese ambivalence toward the accepted narratives and outcomes of World War II.

In a nationalistic reading of the *Gundam* narrative, the statue's initial instalment on Odaiba Island becomes symbolic. The term '*daiba*' means 'fort' in English and Odaiba was built by the Tokugawa shogunate in the mid-nineteenth century, intended as a special man-made island fortress guarding against possible attacks from the sea. The fortification was a response to Perry's gunboat diplomacy in the mid-nineteenth century, which demanded that the Tokugawa government open up Japan's ports for international trade, as I discussed in the introduction. The Gundam statue stood facing the sea, protecting Japan. It is also noteworthy that it was installed less than a kilometre away from a small replica of the Statue of Liberty (of about 12 metres in height) at the Odaiba Marine Park in front of the Aqua City shopping centre, another large shopping mall. This replica was brought to the site in 1998. As it was so popular, a permanent version was installed in 2000. For architectural design theorist and cultural theorist Kaichiro Morikawa, the replica, positioned near the mall, represents Japan's consumerism and its cultural cringe toward the West. Adding to its artificial-attraction, Las Vegas qualities, Odaiba Island is also the home of Venus Fort, a very fashionable shopping mall located perpendicularly to Diver City that is intended to emulate a medieval southern European town (Morikawa 2008, 234–5).

If one sees the Gundam statue as representing the juncture where the nationalistic subtext of the story of *Gundam*, the history of Odaiba Island, and the Westernised cultural landscape of the corporate-led Odaiba leisure precinct meet, the commercial-national 'place' of the statue can be imagined as a battleground of Japan's defence against 'the ruling of American cultural colonialism', according to Morikawa (2008, 3). If Gundam is itself an entirely corporatised showpiece, Morikawa's observation is not entirely accurate: the war has been lost, American-style consumerism has won out, and the only culture that Gundam is protecting is that of Bandai and the shopping mall. The Gundam statue is a blatantly top-down, gimmicky commercial prop. Its framing is overdetermined, allowing limited imaginative engagement for the spectators who are most likely foreign and domestic tourists, young shoppers, or families with children rather than die-hard *Gundam* fans.

In the next section, I examine the manga/anime subculture of the fighting robot and its male fans, and the kind of places created in the enactment of robot fights.

DREAMS OF A GIANT ROBOT

The Japanese manga/anime fans' desire for giant anthropomorphic robots is realised in Kuratas, a piloted humanoid robot developed by iron craftsman and artist Kogoro Kurata. This robot was collaboratively created by Kurata and roboticist Wataru Yoshizaki, who provided the control system for it. Kurata set up Suidobashi Jūkō, a self-funded company that aims to 'mass-produce and sell prototype KURATAS', which is advertised as an art or entertainment piece that 'makes your dream of becoming a robot pilot comes [sic] true' (Kurata 2012). This 4m-high robotic machine weighing 4.5 tons consists of an upper body of humanoid appearance with two arms and four wheel-legs that can extend to lift the body. It is powered by a diesel engine and can move with a maximum speed of 10kph. The machine can be operated with a control device that combines a joystick and steering wheel that are fitted at the pilot's seat in a cockpit inside it. From the cockpit, the outside view is shown on a LCD monitor through cameras. An outside image from a drone can also be transferred to the monitor. The robot can also be operated with an iPhone. Kuratas is fitted with an Xbox Kinect motion sensor that when the pilot smiles, this action triggers 'the smile shot' of twin machine guns that fire 'BBs' (plastic projectile balls) or a rocket launcher that fires water bottles. The base model has a price of JPY 1.2 million (approximately AUD $1.4 million) on Amazon Japan, though this entry is probably in jest, as a buyer is unlikely.

Kuratas debuted in the form of a demonstration presentation at the 2012 Tokyo Wonder Festival, a biannual event for devoted hobbyists who sell and buy models based on popular characters from manga, anime, games, and sci-fi. The festival attracts tens of thousands of people at Makuhari Messe, a mega-scale convention centre in Makuhari New City, a new downtown precinct developed in the land reclamation area outside Tokyo. The Kuratas demonstration was held in a large exhibition hall of bare concrete walls with a high ceiling, like a large army warehouse, and there was a large audience for this event. It started with a promotional video by filmmaker Tadashi Tsukagoshi on a large monitor, and then the machine next to it was unveiled. The MC, Sascha Boeckle, a professional DJ, introduced model Anna Nagae, the female pilot in the promotional video, who would also do the demonstration. Nagae climbed onto the machine and opened the hatch to the cockpit, got inside, and closed the hatch. Soon after the video explaining how to operate Kuratas was screened, Nagae moved Kuratas' arms and twisted the upper body. A drone equipped with a camera flew in from behind. Kuratas pretended to shoot down the drone. The show ended by announcing the sale of Kuratas. Most of the audience members were male hobbyists who were aware of fellow super-hobbyist Kurata, and appreciated his sense of humour, dedication, and craftsmanship. Watching a video from Kurata's website of the event, it is clear that the tone was highly amiable (Kurata 2012).

Kuratas became immediately popular, especially among fans of giant robot anime. Like many of his generation of Japanese born in the early 1970s, from a young age, Kurata was exposed to anime featuring giant robots. Kurata came to believe that Japan, and not other countries, must strive to produce 'workable' giant robots. In 2005, prior to the unveiling of Kuratas, Kurata became known to the robot manga and anime fan community for his creation of a 1:1 scale, four-metre-high static replica of Scopedog, an armoured trooper modelled upon the one that appeared in the 1980s TV anime *Votoms* (Yamanaka 2006). While fans were aware of some small-sized humanoids, such as ASIMO, no-one had yet developed a giant robot. Kurata decided to create Kuratas from scratch. Kurata has indicated that he hopes that the mass production of Kuratas will present a feasible model that will pave the way for 'an age of giant robots' to come (Saijo 2012). Though not an exact replica of a particular robot character, Kuratas suggests the typical mass-produced robotic trooper that appears in robot anime and game culture. Its design also reflects a concern with 'realism' as it is understood in robot anime terms: Kuratas has a grey-coloured metallic body with some wires visible, rather than having a shiny and 'clean' appearance like ASIMO.

Two American engineers, robot enthusiasts Gui Cavalcanti and Matt Oehrlein (MegaBots Inc.), posted a YouTube video in 2015 to challenge Kuratas to a fight with their robot Mk. II. Cavalcanti and Oehrlein's piloted fighting robot is 4.5 m high and 6 tons in weight, requiring two operators. It has a pneumatic big gun that shoots cannonball-sized paintballs at more than 100 km per hour, and a launcher that fires 20 smaller paintballs in a row. Wearing American flags as 'superman's cape', they provocatively state in the video, 'Suidobashi, we have a giant robot. You have a giant robot. You know what needs to happen. We challenge you to a duel' (Cavalcanti and Oehrlein 2015). Kurata accepted the challenge, stating on a reply video that he must win this fight because '[g]iant robots are Japanese culture' (Kurata 2015a). In Kurata's view, the large size of Mk. II and its prominent guns indicate that the robot is really 'super-American' (Kurata 2015a). Kurata proposed a 'melee' style combat, a hand-to-hand fight without projectiles. Both sides are currently negotiating the details of the match (Kurata 2016).

The 'heated' exchanges between Kurata and the MegaBots team indicate a good measure of showmanship and humour about the whole thing, and yet they also appear quite serious and pumped up for the pseudo-fight.

However, there seems to be a slight difference between the two camps in their approach to the event. On the US side, to raise the half million US dollars that would be required for the battle, the MegaBots team launched a crowd-funding campaign and uploaded a promotional video that includes encouraging comments from 'Team America', featuring prominent people and companies within the American robot entertainment and robotics industry, including TV personality Grant Imahara from *Myth Busters*, a popular-science television series, as well as key figures from NASA and other companies working on robots and armoured vehicles (Cavalcanti and Oehrlein 2015). Regardless of the gloss of patriotism voiced in the promotional video, it is obvious that the fight is seen as a possible business opportunity for the television entertainment industry and the robot industry. Indeed, Cavalcanti talks about it in terms of 'Mech sports' (quoted in Casserly 2015). In contrast to the systematised and corporatised plans of the US team, Kurata's approach is not that of a businessman. Though he receives donations, Kurata is not interested in fund-raising, stating that the power of money alone is not enough, and if that were all that mattered, anyone could do it (Kurata 2015b). Kurata remains an 'extreme' hobbyist and craftsman who wants to pursue his vision of the giant robot; he rejects larger schemes that might result in a pragmatic corporatism or his subjection to the requirements of authorities.

For Japanese hobbyists, it is Kurata's eccentric stance that appeals, and Kuratas reflects Kurata's stubborn, hobbyist individualism. In other words, Kuratas the robot, and Kurata the hobbyist, embody certain ideals that attract adult fans of giant robot anime. As I mentioned, the fact that Kuratas does not look slick and factory-produced is attractive to fans. There are some decals on the machine that increase its verisimilitude for fans, as it looks similar to machine weaponry as it is depicted in robot anime. Kuratas is seen as an actualisation of a Gundam-type robot in Japan.

While there are other large, robotic performing machines that can be ridden by a human and are more or less oversized toys for children, such as Sakakibara Kikai Company's Land Walker, which consists of a box-shaped cockpit with two giant legs (Sakakibara Kikai 2016), or Kabutom, a giant robot shaped like a rhinoceros beetle, built by engineer Hitoshi Takahashi (Takahashi 2016), these machines lack factors that make them desirable for adult robot anime enthusiasts: they are childish and not cool in the way that Kuratas is, referencing a hobbyist subculture, an aspect of which involves a sexualisation of the object of interest.

One aspect that appeals to the adult fans of Kuratas is the use of a beautiful, young, female model for Kuratas's promotional video and the demonstration. A reviewer on 'Rocket News 24', a tabloid information site, recognises this eroticising of the robot, commenting that 'the collaboration between a [giant] robot and a beautiful young woman excites the fans of giant robot anime' (2012). In robot manga and anime, plots in which the main character is a young girl who is actually a combat android or cyborg, or is a pilot who operates gigantic robots, is very common; it is an established trope in manga/anime.[3] Toshio Okada, prominent anime producer and commentator on Japanese manga and anime culture, observes that for an anime to be successful, '"All you need is a girl who goes to outer space and a giant robot"' (quoted in Saito 2011a, 5). This combination of giant robotic machine and young, beautiful fighting girl is a potent recipe for the male *otaku* (nerd or geek), a characterisation often used to refer to fans in the manga/anime subculture in Japan. Male *otaku* form a significant part of the fan base for Kuratas's work.

Otaku fans are very particular about their objects of desire. (I will return to *otaku* culture, and the question of desire, in the next chapter.) Sociologist Masachi Osawa calls the male *otaku*'s attitude, in which reality and fiction are regarded in equivalent terms, 'ironical immersion', which suggests the *otaku*'s awareness of his own predilections and his inability to control them (2008, 105). Kuratas is emblematic of Japan's techno-fandom culture in the sense that the modern Japanese idea of technology as central to, and representing, progress, social advancement, economic power, and future success becomes fetishised to the point where it becomes separated from reality, and, in often metaphorically embodied and eroticised forms, generates a different (and preferred) reality. The imagery of fighting robots in manga/anime is often the subject of this fetishisation. While Kurata's response to the MegaBots team reflects the dilemma of the individual's enmeshment in Japan's corporatised and regulated modernity, the fetishisation of the giant robot comes not from Kurata but from the dynamics of Japanese fandom, which is blind to its own appetite for corporate products and fashions, and thus its complicity with the corporatising of even perverse fantasy (a matter I will discuss later on, with regard to the industry that has arisen around the popular Hatsune Miku character).

ROBOT RESTAURANT

The imagery of the Japanese anthropomorphic robot also creates opportunities for the tourist dollar. In contrast to the performance of Gundam for the general Japanese public, and that of Kuratas, which is, in essence, targeting *otaku*, Tokyo's Robot Restaurant draws upon stock images of Japan and its robot culture for the entertainment of foreign tourists. It exploits a more recent narrative, beyond samurai, geisha, and ikebana, of 'the land of the rising sun' through the figure of the robot as a new kind of twenty-first-century performer. Robot Restaurant's cabaret performance for tourists deploys images taken from the vocabulary of foreign fantasies concerning Japan's contemporary technoculture. Based in Shinjuku, an area known for its red-light entertainments, it is not surprising that the restaurant's cabaret acts combine robot imagery and sleaze in an atmosphere of exoticism and erotic charge.

In this regard, Robot Restaurant can be examined along the same lines as Noboru Iguchi's export films *The Machine Girl* (2008) and *Robo Geisha* (2009), in which fighting female cyborgs are the main characters. These films cash in on the imaginary of *otaku* Japan, emphasising the familiar rhetoric of 'techno-Orientalism' (Morley and Robins 1995), which focuses upon the strangeness and exotic nature of the Japanese as the 'other'. Robot Restaurant, providing live entertainment for tourists, similarly mines a techno-Orientalism and, further, structures an economy of 'self-colonisation', in which the colonised participate in—and, at Robot Restaurant, delight in—the fantasies of their colonisers (T. Ueno 2001, 235, original English). Before discussing self-orientalising performance at Robot Restaurant, I first turn to recent studies on tourist consumption that argue for performance-oriented approaches. Tourists present as a particular kind of audience, one that craves transformative experiences and demonstrations of 'authenticity'. Robot Restaurant responds to these desires.

The question of authenticity versus inauthenticity has been prominent in tourism studies. While tourists desire authenticity, they are presented only with 'staged authenticity' (MacCannell 1973). More recent scholarship has nuanced this apparent dichotomy. For example, sociologist Ning Wang identifies different notions of authenticity that are operative in tourist experiences (1999, 350). Objective authenticity is measured by art or science specialists through certain standards, as for artefacts in a museum. Constructive authenticity conceives of authenticity as symbolic and socially constructed, acknowledging 'conflicting perspectives regarding tourist motivations and experience' (Rickly-Boyd 2013, 682). Here, signs of authenticity are appreciated through a projection of stereotypical imagery, expectations, or beliefs by tourists or tour producers onto the objects of tourism (Wang 1999, 355–6). For Wang, considering the postmodern refusal of the original, even 'the contrived, the copy, and [the] imitation' are celebrated (1999, 357). Cultural theorists Michael Haldrup and Jonas Larsen also discuss the relationship between tourism and material culture around the notion of the Orient (2010, 12). According to Haldrup and Larsen, 'the ludic and often ironic circulation of Oriental tropes, clichés and fragments at resorts' should not be taken at face value but regarded as playful, ritualised behaviour that is partly constrained, partly innovative (2010, 99). For cultural theorist Simon Coleman and cultural geographer Mike Crang, at the site of cultural performances for tourists, 'the very idea of authenticity is part of a reflexive poetic and political field – a term to be contested and used' (2002, 7). So, tourism is an engagement with both the 'authentic' and the inauthentic that recognises elements of play, irony, and the contribution of the tourist him- or herself. Cultural anthropologist Yujie Zhu discusses 'the transitional and transformative process' of the tourist experience, stressing 'the dynamic interaction between individual agency and the external world' (2012, 1498).

I deploy a cultural-materialist approach to the relationship between place and performance with regard to the Robot Restaurant's cabaret events, as theatre studies scholar Marvin Carlson discusses it. That is to say I discuss Robot Restaurant in terms of the 'elements of the process by which an audience makes meaning of its experience', through a consideration of 'the entire theatre, its audience arrangements, its other public spaces, its physical appearance, even its location within a city' (Carlson 1989, 2).[4] I take into account the red-light-district location and the elements of both cultural collage and Orientalism of the Robot Restaurant's cabaret show in relation to a history of such entertainments in Tokyo. I also apply to the Robot Restaurant, treating it as a kind of theatre, the expanded notion of the 'theatrical public sphere' discussed by theatre studies scholar Christopher B. Balme, which focuses less on 'the event' itself and more on its 'social and political imbrication' (2014, 14). In other words, I aim to investigate the Robot Restaurant's theatrical public sphere, not merely in terms of its theatre space alone, per Carlson, but in terms of its 'discursive, social and institutional factors' that affect the reception of the show (Balme 2014, 23).

Robot Restaurant opened in July 2012 in Kabukichō, Shinjuku, a seedy entertainment area of Tokyo with movie theatres, host and hostess bars, izakaya bars, night clubs, and shops.[5] According to the director of the club, Namie Osawa, the motivation behind the creation of Robot Restaurant was the rebuilding of the area, which has declined in recent years, via the introduction of a new type of night entertainment combining female dancers with futuristic robots (Yamatogokoro 2013, 2). This venue consists of a few levels in a building called Robot Building (named as such when the restaurant came in). All its interior walls, floors, and pillars are fitted with neon, glitter, flashing lights and LED screens; iridescent, glossy, and reflective surfaces and mirrors; and brightly lit chandeliers. The outlook is ostentatious, over-the-top, and tacky, like a hyper-technologised Las Vegas. It is said that the owner, a successful adult-entertainment businessman, Kei'ichi Morishita, invested a total of JPY 10 billion in the venture (AUD $107.2 million). Robot Restaurant is a sort of cabaret in a theatre on the fourth level down in the underground part of the building. After walking down the psychedelically decorated, narrow stairways, clients are given a bento box and a bottle of cold tea (alcohol can be purchased during breaks) before they are led to rows of stadium seating on both sides of a rectangular performance arena of approximately ten metres by three metres. Its spectacular multimedia cabaret consists of a group of 30 female dancers called Josen (literally meaning 'women fight') and performing robots (rideable androids, exo-skeletal humanoids, and remote-controlled zoomorphic machines), developed, operated, and maintained by a technical crew of 50.[6]

Robot Restaurant's performance—three 1-hour long performances each night—is a multimedia hybrid of a pageant and burlesque show, with dance and music. Large LED panels on the walls are mounted behind the seats, which provide a backdrop of video images for audience members on the other side. Fifteen or so agile young women performers –wearing white or red wigs, or what appear to be New Guinean tribal masks, and wearing kimono, skimpy lingerie, marching band uniforms, or sparkling bikinis— move around the space and dance with animatronic dinosaurs, metallic-costumed robot warriors on roller-blades, Segways, and single-wheeled motorbikes covered with bright neon lights. The dancers perform on a tank and a bomber aircraft made of LED lights as well as on three-and-a-half-metre high 'fembots'. The space is filled with upbeat electronic dance music and laser-beam lights. The performance itself is evolving continuously, adding new robots and themes. It incorporates a wide range of materials, including references to Japanese traditional festivals that use *wadaiko* (traditional drums), Kabuki, sword fighting, dinosaur attacks, marching bands (a reference to the sexy music video for the song *Destination Calabria*), pole-dancing, and robot super-villains, as well as references to the film *Kung-Fu Panda* and the song and video *Gangnam Style*. The audience enjoys a parodic combination of exuberant female bodies, varied forms of actual robotic machines, and a multimedia extravaganza drawing upon an assortment of Japanese and Western pop-cultural references. All of these elements are well organised but presented in a faux-disorderly manner to suggest overabundance and an excitement that exceeds orderliness. At the end of the show, the audience is allowed to take souvenir pictures with the robot performers.

Robot Restaurant trades in 'techno-Orientalism'. The club was initially designed to attract Japanese businessmen, but its opening in July 2012 was immediately covered by the German, American, and British media rather than the local media, and its promotion on the Internet seems to be aimed at foreign customers. When I visited Robot Restaurant in 2013, its customers were overwhelmingly non-Japanese tourists. Osawa proudly indicates that Hollywood sci-fi film directors, such as Tim Burton, J.J. Abrams, and Guillermo del Toro have visited it (Yamatogokoro 2013, 2). Osawa describes the Western tourists favourably as they are easily excited by the show. The following comment from TripAdvisor, an English-language tourist information site, captures why this may be so: it states that the performance offers '[a]ll the crazy stereotypes of Japan in one show' ('Nina T' 2013). The Robot Restaurant's mishmash performance satisfies foreigners' appetite for a technologised (and eroticised), futuristic Japan, and the show includes the commonly seen conjunction of fighting girls and robots in Japanese anime. Japanese tourism has even re-appropriated this techno-fetishistic image: the restaurant's success has been recognised at a government level, and the video footage of the cabaret show is used as part of a promotional website for Shinjuku (Japan National Tourism Organization 2013). The restaurant's potential has been assimilated into the official language of this 'place', one kind of entertainment on offer in red-light Shinjuku.

The locality of Shinjuku has been seen as '*akusho*' (literally meaning 'bad place' or 'evil place'), a pre-modern, Japanese term that is not otherwise in contemporary use. In the past, it referred to marginal places at the edge of a town or city and also metaphorically outside of everyday routine and its spaces (Hiromatsu 1973, 12–3). I see Robot Restaurant as a *geinoh* production, a term I raised in the introduction to this study. Traditionally, *geinoh* practitioners were regarded as 'strangers' who crossed between zones of the sacred and profane, good and bad, inside and outside; they were both settler and floater, in this world and the netherworld; and, in a city, they were forced to live in *akusho* locations (Hiromatsu 1973, 12–3). The fact that Shinjuku is marked as a seedy, contemporary *akusho* in Tokyo lends the restaurant a liminal attractiveness.

The extraterritoriality associated with *akusho* also colours the cultural mishmash of the Robot Restaurant's performances. It is the contemporary version of *Casino Follies,* a cabaret venue in Asakusa that opened in 1929 and had its heyday during the early to mid-1930s. Asakusa was an active entertainment and red-light district in Tokyo of the pre-World War II period, before some entertainments began to move to Ginza. (Asakusa was heavily damaged during the war and was rebuilt.) According to Shunya Yoshimi, Asakusa's role as a red-light centre shifted to Shinjuku in the late 1960s (2008: 277). Historian Miriam Silverberg discusses the '*inchiki*' (phony) performances at *Casino Follies,* and its popular performer Enoken's use of 'incongruities and code switches' between stereotyped, premodern Japanese and Western themes, with gags on rampant capitalism, rapid Westernisation, and modern 'mores' (2006, 235–243). For Silverberg, this Asakusa stage represented 'an instance of the ironic *acharaka* comedy', which relates to her view of 'montage' as an apt metaphor for Japanese modern culture (2006, 239). '*Acharaka*' is an abbreviation of the phrase '*achira kara*' (from over there—that is, from the West). The cultural mix in the performances at *Casino Follies* did not present 'a hybrid culture of East meets West' but was instead 'a jagged montage that clearly revealed its points of conjuncture': for example, Enoken played a samurai committing seppuku (self-disembowelment) to 'the rhythm of jazz' (2006, 239). Silverberg sees the cultural montage of the *Casino Follies* performances as part of an indigenisation of Western influences at the time of Japan's growing confidence as a colonial power (Silverberg 2006, 257). Hence, the gags of *Casino Follies* were targeted to the lives and worlds of a local audience. While the cultural mishmash of Robot Restaurant is also one of montage, the Shinjuku cabaret entertains a foreign tourist audience through its juxtapositions of similarly pitched popular themes calculated to appeal to it. Robot Restaurant sells an exoticised 'here' of contemporary Japan that is non-existent yet provides an 'authentic', and outrageously inauthentic, image of Japan within a touristic economy of representations.

Writer and technology researcher Chris Arkenberg's comments on Robot Restaurant reveal a journalistic recognition of the paradox of the Robot Restaurant—that its representations of Japan are phantasmagoric, but they also capture something of the 'real' Tokyo night that one actually finds outside the door of the cabaret:

> From out of the blanching end-of-show lights, we climb the stairs in a daze, back up the psychedelic kaleidoscope to the chilled winter night of Kabukicho, Shinjuku – so bright and shiny and organic – looking for information in the gaps, over beers and grilled meats and Japanese whisky, lost in translation more than ever but feeling warm and safe and shiny in Tokyo. (2014)

The phenomenological experience and impressions of Robot Restaurant merge into the cityscape of Shinjuku and of Tokyo, described by Arkenberg as a *Blade Runner*-like city.[7]

The techno-Orientalism of the human-machine cabaret at Robot Restaurant is also a 'self-Orientalism', a version of self-colonialism in which it is not only the spectacle of the fantastical and bizarre Far East that foreign visitors wish to see but also the willing participation of the Japanese themselves in such spectacles. For cultural researcher Shoichi Inoue, Japan's self-Orientalism might be a strategy of self-empowerment, mocking the West's desired images of Japan (Shoichi Inoue, quoted in Iwabuchi 2002, 461). But Robot Restaurant is an opportunistic business venture; its intentions are not subversive. The burlesque performance of Robot Restaurant in fact challenges neither the Orientalist view nor the cliché of 'Oriental' female desirability; neither does it question the centrality of the robot in narratives of Japanese modernity, or, more relevantly, in the story of modern Japan that is being offered to tourists. In contrast to the other examples in this chapter, in Robot Restaurant, the symbolic and transformative power of the figure of the robot is animated not for the Japanese but for foreigners. In performance with female dancers and spectacular surroundings, the robot mediates both the modern, Orientalist view of Japan and the exported image of Japan's *otaku* culture through contemporary pop-culture representations of Japan and of the robot.

In the next chapter, I will look at a similar network of meaning in relation to the needs and desires of male *otaku* subculture, the highly mediated and mediating world of the software character Hatsune Miku on the Internet.

Notes

1. The term 'media mix' is used in reference to Japanese media convergence (Steinberg 2012, 6).
2. Tokyo's bid for the 2016 Summer Olympics was unsuccessful, but the city won the subsequent bid for the Olympics in 2020.
3. Examples included android Solty in *SoltyRei* (2005); cyborg Major Motoko Kusanagi in *Ghost in the Shell* (1995); Chise, her body turned into an ultimate robotic weapon, in *Saikano: The Last Love Song on This Little Planet* (2000); and Takaya Noriko, a female high school student who operates Gunbuster, a gigantic robot, in *Toppu wo nerae* (1988), to name but a few. For a discussion of the relationship between robots and girl pilots, see (Yoshida 2004, 98–128).
4. An examination of the importance of site, place, and space in the making of theatre and performance works is an established area of enquiry in theatre and performance studies. Book-length treatments of the topic include Mackintosh (1993), Kaye (2000), Pearson and Shanks (2001), Wiles (2003), Henderson (2004), McAuley (2006), Hill and Paris (2006), McKinnie (2007), Harvie (2009), Pearson (2010), Birch and Tompkins (2012).
5. For the historical development of Shinjuku and its culture, see Yoshimi (2008, 268–94).
6. The club purchased and modified exoskeletal robotic costumes made by the American company King Robota, as well as rideable robotic puppets from the Sakakibara Kikai Co (Matsui 2012).
7. It has been widely discussed that Ridley Scott's experience of Tokyo led to his model of the dystopian Los Angeles of *Blade Runner* (1982). Other Hollywood films such as *Kill Bill: Volume 1* (2003), *Lost in Translation* (2003), and *The Fast and the Furious: Tokyo Drift* (2006) depict the Tokyo experience as a garish, hedonistic dreamscape of desire and nightmares. Japanese studies scholar Steven Brown discusses popular exported animations such as *Akira* (1988) and *Ghost in the Shell* (1995) in terms of their depictions of a futuristic Tokyo inflected by 'cyberpunk' culture and 'posthumanism' (2010).

References

Arkenberg, C. (2014, May 24). Domo Arigato Restaurant Roboto!. *Boing Boing*. http://boingboing.net/2014/05/24/robot.html. Accessed on 12-10-2014.

Balme, C. B. (2014). *The theatrical public sphere*. Cambridge: Cambridge University Press.

Birch, A., & Tompkins, J. (Eds.). (2012). *Performing site-specific theatre: Politics, place, practice*. Basingstoke: Palgrave Macmillan.

Brown, S. T. (2010). *Tokyo cyberpunk: Posthumanism in Japanese visual culture*. New York: Palgrave Macmillan.

Carlson, M. (1989). *Places of performance: The semiotics of theatre architecture*. Ithaca: Cornell University Press.

Casserly, M. (2015, August 20). Giant robots prepare to do battle, as America and Japan go to war. And the US needs YOUR help. *PC advisor*. http://www.pcadvisor.co.uk/news/social-networks/america-vs-japan-in-giant-warrior-robot-battle-help-3619707/. Accessed on 12-10-2015.

Cavalcanti, G., & Oehrlein, M. (2015). The USA vs Japan giant robot duel. *MegaBots*. http://www.megabots.com/. Accessed on 12-10-2015.

Coleman, S., & Crang, M. (2002). Grounded tourists, travelling theory. In S. Coleman & M. Crang (Eds.), *Tourism: Between place and performance* (pp. 1–17). New York: Berghahn Books.

Condry, I. (2013). *The soul of Anime: Collaborative creativity and Japan's media success story*. Durham: Duke University Press.

de Certeau, M. (1984). *The practice of everyday life* (S. Rendall, Trans.). Berkeley: University of California Press.

Endo, M. (2002). *Gandamu, ichinen sensō* [Gundam, One Year's War]. Tokyo: Takarajimasha.

Goodlander, J. (2015). Plaza Indonesia: Performing modernity in a shopping mall. In M. Omasta & D. Chappell (Eds.), *Play, performance, and identity: How institutions structure ludic spaces* (pp. 117–127). New York: Routledge.

Haldrup, M., & Larsen, J. (2010). *Tourism, performance and the everyday: Consuming the orient*. London/New York: Routledge.

Harvie, J. (2009). *Theatre and the city*. Basingstoke/New York: Palgrave Macmillan.

Henderson, M. C. (2004). *The city and the theatre: The history of New York playhouses: A 250 year journey from Bowling Green to Times Square*. New York: Back Stage Books.

Hill, L., & Paris, H. (2006). *Performance and place*. Houndmills/Basingstoke/Hampshire/New York: Palgrave Macmillan.

Hiromatsu, T. (1973). *Henkai no akusho* [Bad places at the margins of society]. Tokyo: Heibonsha.

IT Media News. (2009, September 1). Jitsubutsudai Gandamu, kaitai stāto, raijōsha wa sanbai [Actual sized Gundam, started to be dismantled, (results in) three times as many visitors]. http://www.itmedia.co.jp/news/articles/0909/01/news077.html. Accessed on 03-15-2012.

Iwabuchi, K. (2002). 'Soft' nationalism and narcissism: Japanese popular culture goes global. *Asian Studies Review, 26*(4), 447–469. doi:10.1080/ 10357820208713357.

Japan National Tourism Organization. (2013). A city with two faces. *Discover the Sprit of Japan.* http://www.visitjapan.jp/en/m/player/?video=72. Accessed on 12-2-2013.

Jenkins, H. (2006). *Convergence culture: Where old and new media collide.* New York: New York University Press.

Kaye, N. (2000). *Site-specific art performance, place, and documentation.* London/New York: Routledge.

Klingmann, A. (2007). *Brandscapes: Architecture in the experience economy.* Cambridge, MA: MIT Press.

Kurata, K. (2012). KURATAS. *Suidobashi Jūkō.* http://suidobashijuko.jp/index.php. Accessed on 03-10-2014.

Kurata, K. (2015a). Response to robot duel challenge. *Suidobashi Jūkō.* http:// suidobashijuko.jp/index.php. Accessed on 12-10-2015.

Kurata, K. (2015b, July 23). Kuratasu vs Megabotto, sono 2 [Kuratas vs MegaBot, Part 2]. Nandemo tsukuruyo: Hontoni ugokuka kyodai robo [I would make anything: Does the giant robot really move?] (blog). http://monkeyfarm. cocolog-nifty.com/nandemo/2015/07/post-5853.html. Accessed on 10-08-2015.

Kurata, K. (2016, September 6). Kinkyō houkoku nazo [Latest News]. Nandemo tsukuruyo: Hontoni ugokuka kyodai robo [I Would Make Anything: Does the Giant Robot Really Move?] (blog). http://monkeyfarm.cocolog-nifty.com/ nandemo/2016/09/post-31bf.html. Accessed on 20-10-2016.

Lash, S., & Lury, C. (2007). *Global culture industry: The mediation of things.* Cambridge: Polity.

MacCannell, D. (1973). Staged authenticity: Arrangements of social space in tourist settings. *American Journal of Sociology, 79*(3), 589–603.

Mackintosh, I. (1993). *Architecture, actor, and audience.* London/New York: Routledge.

Matsui, Y. (2012, October 5). Futuristic bot cabaret wows Tokyo. *The Japan Times Online.* http://www.japantimes.co.jp/news/2012/10/05/national/futuristic-bot-cabaret-wows-tokyo/. Accessed on 01-15-2013.

McAuley, G. (Ed.). (2006). *Unstable ground: Performance and the politics of place.* Bruxelles/Oxford: P.I.E. Peter Lang.

McKinnie, M. (2007). *City stages: Theatre and urban space in a global city.* Toronto: University of Toronto Press.

Miles, M. (1997). *Art, space and the city: Public art and urban futures.* London/ New York: Routledge.

Morikawa, K. (2008). *Shuto no tanjō: Moeru toshi Akihabara, zōhoban* [The birth of Hobby city: 'Moe' city Akihabara, an expanded edition]. Tokyo: Gentōsha.

Morley, D., & Robins, K. (1995). *Space of identity: Global media, electronic landscapes and cultural boundaries.* London/New York: Routledge.

Nina, T. (2013, December 12). All the crazy stereotypes of Japan in one show. Comments on TripAdvisor. *Robot Restaurant.* http://www.tripadvisor.com.au/Attraction_Review-g1066457-d4776370-Reviews-or20-Robot_Restaurant-Shinjuku_Tokyo_Tokyo_Prefecture_Kanto.html#REVIEWS. Accessed on 12-20-2013.

Osawa, M. (2008). *Fukanōsei no jidai* [*The age of impossibility*]. Tokyo: Iwanami Shoten.

Pearson, M. (2010). *Site-specific performance.* Houndmills/Basingstoke/Hampshire/New York: Palgrave Macmillan.

Pearson, M., & Shanks, M. (2001). *Theatre/archaeology.* London/New York: Routledge.

Rickly-Boyd, J. M. (2013). Existential authenticity: Place matters. *Tourism Geographies, 15*(4), 680–686. doi:10.1080/14616688.2012.762691.

Saijo, T. (2012, July). Hito ga notte sōjū dekiru kyodai robotto, Kuratasu [Kuratas, A giant robot that can be piloted]. *Wired Japan.* http://wired.jp/2012/07/26/kuratas/. Accessed on 12-15-2013.

Saito, T. (2011a). *Beautiful fighting girl* (J. K. Vincent & D. Lawson, Trans.). Minneapolis: The University of Minnesota Press.

Sakakibara Kikai. (2016). Landwalker. http://www.sakakibara-kikai.co.jp/products/other/LW.htm. Accessed on 30-01-2016.

Silverberg, M. R. (2006). *Erotic grotesque nonsense the mass culture of Japanese modern times.* Berkeley: University of California Press.

Stalker, D., & Glymour, C. (1982). The malignant object: Thoughts on public sculpture. *The Public Interest, 66,* 3–21.

Steinberg, M. (2012). *Anime's media mix franchising toys and characters in Japan.* Minneapolis: University of Minnesota Press.

Takahashi, H. (2016). KABUTOM RX-3: Beetle robot official HP. http://kabutom.com/. Accessed on 20-01-2016.

Tane, K. (2010). *Gandamu to nihonjin* [Gundam and the Japanese]. Tokyo: Bungē shunjū.

Ueno, T. (2001). Japanimation and techno-orientalism. In G. Bruce (Ed.), *The uncanny: Experiments in cyborg culture* (pp. 223–231). Vancouver: Vancouver Art Gallery and Arsenal Pulp Press.

Wang, N. (1999). Rethinking authenticity in tourism experience. *Annals of Tourism Research, 26*(2), 349–370. doi:10.1016/S0160-7383(98)00103-0.

Wiles, D. (2003). *A short history of Western performance space.* New York: Cambridge University Press.

Yamanaka, H. (2006). Botomuzu wo tsukutte shimatta otoko, kataru, Part 1 [A Talk by the Man who Built Votoms, Part 1]. *NBOnline Premium.* http://business.nikkeibp.co.jp/free/x/20060328/20060328005467.shtml. Accessed on 10-01-2014.

Yamatogokoro. (2013). Kabukichō no robotto resutoran ni naze gaikoku kyaku ga afureteirunoka? [Why does robot restaurant in Kabukicho attract so many foreign tourists?] *Yamatogokoro.jp.* http://www.yamatogokoro.jp/inbound-interview/index06.html. Accessed on 12-15-2013.

Yoshida, M. (2004). *Nijigen bishōjo ron* [A theory of the two-dimensional beautiful girl]. Tokyo: Futami Shobō.

Yoshimi, S. (2008). *Toshi no doramatrugī : Tokyo sakariba no rekishi* [Dramaturgy of the city: History of entertainment districts in Tokyo]. Tokyo: Kawade shobō.

Zhu, Y. (2012). Performing heritage: Rethinking authenticity in tourism. *Annals of Tourism Research, 39*(3), 1495–1513. doi:10.1016/j.annals.2012.04.003.

CHAPTER 6

Hatsune Miku, Virtual Machine-Woman

This chapter examines the virtual performances of Hatsune Miku, commercially produced software that combines synthesised singing with an illustrated girl character as the singer. This character's name means 'the first sound from the future'. Users of this software create desktop song performances with interpreted images of Hatsune Miku and post them onto Nico Nico Dōga, a popular video-sharing site in Japan. This creation and posting of videos of the singing Hatsune Miku became an Internet sensation immediately after the product's release in 2007. Since then, a wide range of content derivatives has arisen in the forms of illustrations, animation, games, and 3D figures. Fans have also recorded imitative dancing and singing performances of Hatsune Miku, and posted them on the Internet. Hatsune Miku songs have been popular among the younger generations of karaoke fans. There have even been successful 'live' performances featuring the projection of an animated Hatsune Miku.[1]

Hatsune Miku appeals to both male and female fans in Japan, resulting in very diverse creations. Its popularity has been discussed as being due to a complex combination of social and cultural factors, including the terms of fandom in Japan, manga/anime, the conventions of '*kawaii*' (cuteness), and the phenomenon of the idol in the Japanese popular entertainment industry. While discussing these interrelated themes, I will critically examine Hatsune Miku from a feminist perspective, focusing specifically on the generations of male *otaku* (nerd or geek) born after 1970. Fanatic male fans in these groups are said to have *moe* (intense affection, including sexual desire) toward young female characters in manga/anime and games.[2] The *moe* generations of *otaku*

© The Author(s) 2017
Y. Sone, *Japanese Robot Culture*,
DOI 10.1057/978-1-137-52527-7_6

fans of Hatsune Miku manipulate images of their virtual diva. While Pygmalion in ancient Greek mythology falls in love with a sculpture of a woman that he has carved, male *otaku* fall in love with 2D or 3D images of virtual idols they create, or with others' creations on a computer screen. I discuss Hatsune Miku as a representation of a Pygmalion-like desire related to *otaku* subculture, which is in turn an offshoot of the more mainstream Japanese manga/ anime culture.[3] Hatsune Miku is a symbol of an unattainable ideal woman with presumed power over men.

Through an examination of the *otaku*'s engagement with Hatsune Miku, this chapter highlights contemporary sexual politics in Japan in which women are seen in relation to gender stereotypes, pivoted around the ideal of the '*shōjo*' (girl). According to critic Eiji Otsuka, the category of *shōjo* was created when the Meiji government introduced a restriction on underage marriage. In Japan's pre-modern era, there were only two types of woman: sexually immature before menarche, and mature and therefore fertile (1997, 18). *Shōjo* is a term for a woman in a state between menarche and the legal age of marriage. During this time, the girl was to remain virginal, and would acquire education, thereby adding value to her worth for the time when marriage was to be arranged between families (E. Otsuka 1997, 18). The *shōjo* was therefore intended to be unobtainable. Hatsune Miku is supposed to be 16 years old, as I discuss shortly; she is portrayed as wholesome and innocent, and by virtue of her virtual form, she is unreachable and therefore desirable, a contemporary version of the *shōjo*, a product of Meiji modernisation.

This chapter focuses on the manifestations of *moe* practice with regard to Hatsune Miku, that is, a 'technological consumption' of the signs of femininity as another instance of Japanese 'robo-sexism', as in Jennifer Robertson's discussion (2010). Robertson critiques how Japanese roboticists reproduce Japanese gender stereotypes, not least in terms of the appearance of their 'female' humanoids and androids, which usually appear busty, slim, or sexy. Issues of gender stereotyping would become undeniable if and when companion robots or even 'sex-bot' robots were ever built.[4] In the current situation, such a 'gynoid' exists only in the realm of the imagination. While Robertson examines the producers, that is, Japanese roboticists, of gendered images of the humanoid and android, this chapter looks into 'robo-sexism' from the receivers' side—that is, from the perspective of male *otaku* fans who are conscious of their own desire for Hatsune Miku.

The nature of 'robo-sexism' surrounding the *moe* consumption of Hatsune Miku is an effect of the intersection of several different fandom cultures in Japan. I will first discuss the genesis of Hatsune Miku. Then I locate the *moe* that is felt for Hatsune Miku in its larger context, linking it with the cultures of *otaku* fandom, '*lolicon*' manga (a genre of Japanese comics that depicts sexualised, young girl characters), and the female 'idol' (young women pop stars) and relate this last to past attempts to create a 'virtual idol', a female celebrity in the form of animation or computer-generated imagery.

THE DEVELOPMENT OF HATSUNE MIKU

Hatsune Miku is a commercial software product that synthesises a female singing voice. A user of Hatsune Miku inputs lyrics and a melody to make a song. The software was released in 2007 by Crypton Future Media, a media company founded by Hiroyuki Ito in 1995, which develops and sells products for music such as sound effects, mobile phone content, and sound generation software. It is based on Yamaha Corporation's computer music software engine called Vocaloid 2, which uses voice actor Saki Fujita's voice. Ito explains that the reason for selecting the voice of Saki Fujita is because she possesses a natural 'Lolita voice', meaning an adult female actor's childlike voice that is typically used in animation. This voice in itself is understood to express a message of cuteness, which is a central attribute of stereotypical Japanese femininity (quoted in Shiba 2014, 104–5). Prior to Hatsune Miku, Crypton Future Media produced Meiko (based on the voice of Meiko Haigo, a singer songwriter) in 2004, followed by Kaito (the voice of Naoto Fuga, a studio musician) in 2006, using Vocaloid 1. The use of a cartoon image for these products was a promotional strategy to generate a sense that there is a human being 'inside' who is singing. According to Ito, Meiko sold three thousand packages in the first year instead of one thousand, which is the average yearly sales figure for software of this kind (interviewed in T. Kubo 2011). Sensing a possible market for another product after Meiko and Kaito, Ito launched Hatsune Miku with a much more sophisticated image that was designed by a professional illustrator, KEI. While the images of Meiko and Kaito are more like manga images of the past, the illustration of Hatsune Miku looks more contemporary to the Japanese audience. Hatsune Miku is a slender, teenaged girl with long, turquoise twin ponytails, with a typical anime face that features large eyes.

Ito conceived Hatsune Miku as the first product of the 'Character Vocal Series'. To generate a sense of this character's 'real' existence, a few attributes were specified. For example, Hatsune Miku is stated as being 16 years old, and her height and weight are set as 158 cm and 42 kg. Hatsune Miku is 'good at' pop music and dance music (Crypton Future Media 2016b).

Hatsune Miku sold more than forty thousand units after a year. The most notable marker of its success is that a compilation CD album of Vocaloid songs, featuring Hatsune Miku songs as its main component, was ranked as the number-one bestseller in May 2010. Ito continues to produce products for the Character Vocal Series: Kagamine Rin and Len, with voice actor Asami Shimoda's voice (2007), are put forward as twin brother and sister aged 14 years old. Megurine Luka (2009), based on Voice actor Yu Asakawa's voice, is set slightly older, at 20 years old. An upgraded version of Kaito has been released in 2013 with Vocaloid 3. A new version of Hatsune Miku using Vocaloid 4 was released in the summer of 2016 (Crypton Future Media 2016a).

Hatsune Miku's popularity reflects Ito's business strategy for the Character Vocal Series work. Ito is an advocate of consumer-generated media (CGM) (T. Kubo 2011). This term refers to content created by consumers and users of commercial products that is published by them on social media platforms. He believes that user-generated content (UGC) is what develops his business. Otsuka points out that copyright holders—of manga or animation, for example—tolerate secondary creations and associated copyright infringement because these illegal activities can create a new market for the products whose copyright is supposedly violated (2012, 13). In his business plan for Hatsune Miku, therefore, Ito allows the secondary generation of music pieces and images by non-commercial users, and these may involve his company's music sources and illustrations of characters in the Character Vocal Series, contrary to the standard practice within the music industry, which lays down rules to protect copyrighted materials. Users can freely create their own song music, combining it with the original image of Hatsune Miku or his or her own interpretation of the image. In order to further facilitate user activities, Ito also created Piapro (an abbreviation of peer production), a website where its members can exchange their own works of music, illustration, and song lyrics. Users are allowed to use other users' works, whether music, illustration, song lyrics, or 3D models, as long as the authors' names are acknowledged. Collaborative activities between users are also encouraged. For example, a user of the site who uploaded a secondary image of Hatsune Miku may produce a video clip with someone who provides original song music with another user's song lyrics (I will discuss this kind of sharing activity within Japanese fandom subculture shortly).

Hatsune Miku became significantly more popular through Nico Nico Dōga, a popular video-sharing site in Japan that is like YouTube. It began its operation in 2006, a year before the release of Hatsune Miku. This site facilitates successive creations by users of songs, illustrations, and videos that relate to Hatsune Miku. Just like Ito's Piapro, members are allowed to use others' works with permission and acknowledgement. Unlike YouTube, however, viewers can insert comments on any video clips they are viewing, creating a sense of 'live' participation. For example, if a viewer inserts the comment 'Wow' via the website's message tool onto an uploaded video clip, the word will appear, just like subtitles do on the television screen, running across the video image. While most of the texts on video clips are interjectory expressions such as 'great' or 'excellent', there are also short comments, such as 'I like this song' or 'this Hatsune Miku is very cute'. In this way, viewers provide instant feedback on video clips. The more popular a clip gets, the more messages the screen is covered with. If there are many messages at a particular section of the video clip, they are automatically sorted in rows and stream one after the other. A user's message is shown together with others' comments. When it is played back, the comments across the screen generate a sense that one is watching the video clip communally with others, just like watching TV with others or going to the cinema (Hamano 2008, 229).

The other important development within the Hatsune Miku 'infrastructure' was the release of MMD (MikuMikuDance) by Yu Higuchi in 2008. It is freeware that allows users to create animation movies of the Hatsune Miku character. The users of MMD create animations of the Hatsune Miku character that sing songs just like Japanese pop singers. Hatsune Miku gained its singing and dancing 'body' via MMD. The animations of Hatsune Miku's singing performances were, and continue to be, uploaded onto Nico Nico Dōga, establishing a popular genre within the website. There were regular competitions two or three times a year to judge the artistry of fans' animations. As well, the first staging of a live concert, which involved actual attendance by fans and featured projected animation of the character through MMD, was organised in 2009. Based on the success of the first concert, game company Sega decided to produce another 'live' concert in March of the following year. It was conceived as a Hatsune Miku promotion for Project DIVA, a rhythm game with 3D images, which was launched by Sega and Crypton Future Media in 2009. The image of Hatsune Miku was the CG image from the game. 'Live' concerts of Hatsune Miku became annual events. These concerts attract thousands of Hatsune Miku fans. The performances have been simultaneously live-streamed. A range of Hatsune Miku merchandise is available at these concerts, including Blu-ray and DVD recordings, as well as small figurines of Hatsune Miku, a collectable item made by 'Nendoroid', a popular brand of small plastic figure replicas of manga and anime characters. Digests of Hatsune Miku concerts are uploaded to the Nico Nico Dōga website. These clips, seen on the website with viewers' messages, create a secondary, delayed 'live' concert for the audience. The clips with scrolling texts present the visualised, collective 'cheering voice' of its viewers, as if they were theatre spectators (Hamano 2008, 218).

Hatsune Miku exists as a network of products across different media platforms, and together they assume an affective power for fans. The Hatsune Miku boom points to the complex interaction between consumer culture, popular culture, and *otaku* subculture. Before returning to Hatsune Miku, in the section below, I discuss the history of the comics market, *otaku* culture, and *Lolicon* manga (a genre of manga that features the images of young girls).

Japanese Fanzine Culture and *Otaku*

Ito's argument for CGM needs to be understood in relation to the Japanese 'fanzine' (or 'zine') culture. The modern form of fanzine activity in Japan began in the Meiji period, when aspiring writers published their novels and poems in *dōjinshi*, self-published magazines. After World War II, emerging manga and illustration artists published their original works through *dōjinshi*, just as writers in previous generations had done. Since the 1950s, new works using or based on known animation and manga have become much more popular among fans and hobbyists, and many *dōjinshi* specialising in secondary creations, like 'slash' novels and manga redeploying established characters, have been developed. These creative activities can be correlated to their Western counterpart, the products of what media theorist John Fiske defines as a fan who is 'an "excessive reader" who differs from the "ordinary" one in degree rather than kind', someone who reads into the work, inventing '"extra-textual" relationships' between characters in popular films or television dramas, such as *Star Trek* (Fiske 1992, 46). Since 1975, there has been a regular comics market ('*komike*' or '*komiketto*' in Japanese) in Tokyo, and secondary creations are also sold there. The comics market has greatly increased fanzine readership, as well as establishing a trade in secondary works.[5] The market has become a biannual event, attracting huge crowds of more than 500,000 in recent years.

The Japanese fanzine culture in the pre-Internet computer age included secondary creations through tapes, videos known as 'MAD' (spelled always in capitals and meaning 'crazy') movies, 'MAD' anime, and *dōjin ongaku* (secondary musical compositions), which then migrated onto Internet sites. These practices are appropriations of film, television footage, animation, and music that might be described as assemblage, collage, cut up, remix, or mash-up. The secondary creations that result are often regarded as infringing copyright in Japan. By creating a platform that facilitates these secondary creations, Ito aimed to channel the creative desires of fans toward Hatsune Miku, inspired by Creative Commons.

I would like to explain the term *otaku*—which is used to refer to a subgroup of fans in Japanese fan culture—before I relate the *otaku* to Hatsune Miku. It attracts much attention in discussions of Japanese popular culture, consumerism, and technoculture.

The literal meaning of *otaku* is 'your house', and it can be used to address 'you' in a formal manner. It was a slang word used among male fans and hobby enthusiasts in the 1980s to describe themselves in a self-deprecating manner. It was meant as a 'witty reference' to male hobbyists' stilted communication style, as they would communicate with their peers using a 'distant and over-formal form of address' because of their clumsy interpersonal communication skills and reclusive natures (Kinsella 2000, 128). The term became publically known when critic and *dōjinshi* creator Akio Nakajima used the term to criticise the male-nerd type of character that appeared in his series of writings in the adult manga magazine *Manga Burikko* in 1983 (Yoshimoto 2009, 172–3). The term was received extremely negatively in the late 1980s when Tsutomu Miyazaki, a paedophile serial killer, was described in the public domain as an *otaku*. Hence, the male *otaku* hobbyist or fan was typically regarded negatively in the late 1980s and the early 1990s because of his presumed antisocial attitudes and dysfunctional behaviour; the *otaku* also became a symbol of alienation among Japanese youth (Kinsella 2000, 136–7).

While the image of the *otaku* still retains negative connotations, it has become, in the twenty-first century, a complex and elusive term that addresses varied practices and fandom-related activities. As part of global strategies to promote Japan, *otaku* culture has become an important, grassroots element of Japanese 'soft power' (Iwabuchi 2002). *Otaku* culture has been exported overseas, and, according to cultural anthropologist Mizuko Ito, it is 'situated at a transnational confluence of social, cultural, and technological trends' that contribute to shaping technology, media, and entertainment (2012, xii).

The targets of *otaku* fans' interest can include manga, anime, games, the Internet, computers, books, figurines, celebrities, special effects, and cosplay (costume play) (Enomoto 2009). While *otaku* is roughly equivalent to the English term 'nerd' and usually refers to men, it can also be used for women as in *onna* (woman) *otaku*.[6] The stereotype of the male *otaku* is that of an introvert, but *otaku* who frequent game arcades, on the other hand, are thought to be extroverts (Kijima 2012, 249–74). *Otaku* fandom activities can also facilitate a collective social bond among members of a particular *otaku* subculture (Stevens 2010, 212).

While acknowledging the complexity of *otaku* culture, I will focus on the most controversial aspect of Japanese male *otaku* practice, which is to do with sexual fantasy concerning cartoon images of young girl characters. Die-hard male *otaku* of the *moe* generations are known to admire cartoon images of young girl characters, particularly in the context of fan subcultures, and can develop strong psychosexual bonds with these characters. These *otaku* males prefer signs of girlishness as expressed in manga/anime over dealings with a real girl. For Thomas LaMarre, a scholar on Japanese animation, it is 'a world of male masturbatory pleasure, a world of sex without actual women' (2009, 241). Japanese literary studies scholar Keith Vincent, on the other hand, discusses the Japanese view of the male *otaku*'s sexual investment as 'perversion' (2011, xviii). Referring to queer theorists such as Eve Kosofsky Sedgwick, Judith Butler, and Michael Moon, Vincent resituates the male *otaku*'s 'perversion' as queer within the Japanese context, highlighting the *otaku*'s denial of distinctions between sexual fantasy and reality; his separation of sexual desire, social identities, and natural bodies; and his implicit stance against the heterosexual normativity of sexual relations involving real bodies (2011, xviii–xxii). However, the *otaku*'s rejection of these aspects, including heteronormativity, is not exactly the same as 'queer', as it is understood in the West, as a sexual-political term. The *otaku* investment, and its separation from social and sexual norms for adult behaviour, is better described by the Japanese term *hentai*, which retains a sense of social derogation, though in Japan it is often the word chosen to translate 'queer'.[7] *Hentai* signifies a wide range of sexually 'deviant', 'perverse', and 'abnormal' practices as seen from the perspective of mainstream, heterosexual normalcy in Japan, and it also reflects a rejection of heteronormativity, as does 'queer'. I feel this term is more relevant than 'queer' to indicate such Japanese subcultural practices that are, importantly, not intended as critical and, though they remain in the realm of the imagination and are not actualised, can involve cartoon images of young girls who would, in real life, be considered underage.[8]

LOLICON MANGA AND OTAKU DESIRE

At comics markets, pornographic parodies of well-known manga/anime are enormously popular among the *otaku* fans who collect such works.[9] They are often categorised as one type of *lolicon* (an abbreviation of 'Lolita complex', sometimes spelled as *rorikon*) manga, a genre of manga that expresses 'the desire for fictional girl characters', rather than 'the desire of an older man for a young girl', as the Lolita reference is usually understood, after the eponymous film and the novel by Vladimir Nabokov (Galbraith 2015, 208).[10] Cultural critic Kazuko Nimiya explains that fans of *lolicon* are attracted to an image of a woman with a childlike face that projects a sense of innocence and purity coupled with a voluptuous and sexualised body (2009, 310). Manga artist Hideo Azuma is regarded as the key figure for the development of *lolicon* imagery. The publication of fanzine '*Cybele*' at *comike 11* in 1979 by Azuma, with Yukao Oki as editor, is regarded within *otaku* subculture as a benchmark event in the history of *lolicon* manga (Morikawa 2011, 180). This anthology of erotic manga that depicted sexual acts by young girl cartoon characters triggered a boom for *lolicon* fanzines in the early 1980s (Sasakibara 2004, 36). Azuma and Oki felt that there was a need to counter the mainstream manga that appealed to a female readership, and they also wished to express their dissatisfaction with *ero gekiga* (realistic erotic manga magazines) for male readers (Morikawa 2011, 181). Azuma was the first professional manga artist who consciously produced parodic, pornographic *lolicon* manga in the style of Osamu Tezuka in commercial magazines of the early 1980s (E. Otsuka 2001, 93; Sasakibara 2004, 38). Otsuka explains that Azuma took Tezuka's cute manga characters, simple line drawings that nonetheless expressed their interiority, and sexualised them, using Tezuka's visual language (2009, 197).

For Azuma, the *lolicon* manga presented a parody of Tezuka's manga, which may have been mildly suggestive of sexuality but never explicitly revealed it (E. Otsuka and Nobuaki Osawa 2005, 170). Azuma's work was intended as a parody, and Tezuka's characters were sexualised with parody in mind; his *lolicon* manga were not meant to be masturbation material (Takatsuki 2009, 104). Nimiya points out that while the *lolicon* manga fans of the 1980s and 1990s are more attracted to innocence and shyness in depictions of young girl characters, the younger generations of *otaku*, those who experience *moe*, the intense excitement and desire that drives them to consume these images, prefer exposure of young girl characters' bodies (2009, 310–11). How can the latter reception be understood?

Lolicon manga and *otaku* sexuality has been discussed through semiological studies on Japanese manga. It is understood that what male *otaku* respond to is in fact the signs of sexiness. In other words, the *otaku* care about *ideas* of sexiness. Male *otaku* are, then, responding to ideas of the erotic that are generated through these manga/anime images of young women. Images valued by male *otaku* in the 1990s were not based on particular works or images by particular artists but instead represented a synthesis of fragments and parts from what philosopher Hiroki Azuma calls the 'database' of this culturally specific imaginary, including manga, anime, and SF films alongside more traditional, iconic images within Japanese visual culture (2001, 52). However, these syntheses cannot be characterised by free association, because there are particular codes and styles through which the *otaku* feels *moe* (Azuma 2001, 66–70). Part of the *otaku*'s approach to images includes '*moe gijinka*' (*moe* anthropomorphism), through which nonhuman entities (physical or virtual objects, commercial products, or even concepts) are anthropomorphised into beautiful young girls. For Go Ito, *moe* can be a response to the intensity of a *kyara*, meaning a cartoon character that assumes some sort of symbolic or iconic power (2005, 104). It is along these lines that psychiatrist and critic Tamaki Saito discusses *otaku* sexual fantasies as *kyara moe* (2011a, 142). Male *otaku* are, then, responding to ideas of the erotic that are metonymically generated through these images.

Hatsune Miku is an *otaku* icon of the young girl, plus the technology-related themes that it calls upon, both of which draw *otaku* interest. For example, Akinori Kubo discusses Hatsune Miku's appeal in terms of the general lack of interiority of *lolicon* manga, concerned with images of young girls as signs, and manga/anime tropes of the controllable, rideable robot character and the networked cyborg body. While the *lolicon* manga *kyara* are understood as artificial constructs with no interiority, the male fans project lust and love onto these two-dimensional images (A. Kubo 2015, 99). Male manga/anime fans appreciate *Iron Man No 28* or *Mazinger Z* from the pilot's viewpoint of controlling a destructive machine that has superhuman powers (A. Kubo 2015, 87). A genre within *otaku* iconography combines the categories of *shōjo* and machine: *meka bishōjo* (mechanical pretty girl), also known as *meka shōjo* (mechanic girl) or *robo musume* (robot girl). These characters embody *lolicon* manga *kyara*, combined with weapons drawn from robot manga characters such as Gundam or presented as girl-next-door-yet-super-android, satisfying the *otaku*'s dislike for the real body of a woman and their attraction to controllable machines (Yoshida 2004, 73). On the cyborg aspect, Kubo links Hatsune Miku with Major Motoko Kusanagi, a cyborg that is the main character of the popular science fiction manga/anime *Ghost in the Shell*, who embodies an ambivalent subjectivity that links and is linked to (or hacks and is hacked) by others through the Net (2015, 104).[11] These tropes combine to suggest, for the Kusanagi character, the blankness of a female machine body that can be controlled and ridden. Kubo argues that Hatsune Miku similarly functions as an empty vessel, inviting and linking countless users: Hatsune Miku is a composite of the hollowness of the sign, the instrumentality of technology, and the innocent girl image, which together invite male fans to fill her with meaning (2015, 110).

While Kubo highlights the pervasive effects of sign and image in *otaku* culture, he treats the blatantly sexist thrust of the mechanical girl character in manga/anime in neutral terms and does not question it or critically evaluate its social and political implications. I will bring a critical perspective to my examination of *otaku* engagement with Hatsune Miku by first examining the idea of the mechanical woman and then, in the following section, the culture of the female idol.

CRITIQUE OF THE *OTAKU* IMAGINATION

I will first refer to critical readings by Western scholars of narratives of artificially created female characters: Olympia in E.T.A. Hoffmann's *The Sandman* (1816); Hadaly in Auguste Villiers de l'Isle-Adam's *Tomorrow's Eve* (1886); and robot Maria in Fritz Lang's 1927 film, *Metropolis*. These stories are informed by speculation on or reaction to changing social relations as a result of far-reaching political and technological change in the course of Western modernity, and a related fear of woman and of the lower classes, and thus the desire to control her/them. Such changes include not only the steam engine, the electric light bulb, and the automobile but events such as women's suffrage and the Bolshevik Revolution. In the stories of Olympia and Hadaly, technology and woman are conflated and regarded as dangerous, possessing destructive power, and located 'in the position of the other, unknowable' (Miller Frank 1995, 161). Comparative literature scholar Andreas Huyssen explains the allure of robot Maria: 'the machine-woman, who is no longer recognized as a machine, makes all men lose control' (1986, 77). Onto the body of robot Maria, according to Huyssen, fears and anxieties about machines and technology are 'recast and reconstructed in terms of the male fear of female sexuality', inducing 'castration anxiety' (1986, 70). The machine-woman is also the idealised image of woman, like Pygmalion's statue; it is a man-made replacement of woman. The machine-woman symbolises technology that offers the means to address woman's supposed inferior nature and represents man's desire for control and possession not only of the opposite sex but also for a harnessing of the power of creation, previously that of woman alone (1986, 71). That the wish to recreate an ideal woman through technological means is fraught with anxiety about both woman and technology also characterises the Japanese version of this narrative and its contemporary iteration in images of woman in popular culture, such as *otaku* culture, *lolicon* manga, and Hatsune Miku.

For example, as I discussed in previous chapters, Japanese (male) fear, anxiety, or ambivalence toward technologies resonates throughout the history of Japan's encounters with modernity: one need only think of major historical events such as Perry's gunboat diplomacy, the defeat in World War II, and the atomic bomb. A general disillusionment toward science and technology—antithesis to the view of science and robotics in particular as guaranteeing a bright future for Japan—set in during Japan's economic decline and stagnation from the 1990s onward, as I have mentioned earlier, and as Yanobe's artwork illustrates. As for anxiety concerning women and the relationship between women's position in post-war Japanese society and *lolicon* manga, Japanese Studies scholar Sharon Kinsella summarises that the male fans of *lolicon* manga, who are unable to relate to 'assertive and insubordinate contemporary young women', are aware of 'the increasing power and centrality of young women in society, as well as a reactive desire to see these young women infantilized, undressed and subordinate' (2000, 122).[12] Otsuka points out that Hideo Azuma's *lolicon* manga of the 1980s portray men's inability to engage with the body of a real woman and the 'otherness' of woman (2001, 98–101). For manga critic Go Sasakibara, the format of *bishōjo* (pretty young girl) games (otherwise known as *gyarugē*) since the 1990s tends to commingle both pornographic and romantic narratives, presenting a guilt-free space for male users to consume young girl cartoon characters (2004, 182). For Otsuka and social critic Nobuaki Osawa, practices of *moe*, which are merely pornographic and lack the potentially critical element of parody, re-establish a patriarchal structure in which men control young girls as their possessions (2005, 170).

Social commenter and editor Tsunehiro Uno, on the other hand, discusses the disenchantment of the 1990s youth, highlighting the year 1995 as a 'turning point' in recent Japanese history: as I have mentioned, this was the year of the Great Hanshin Earthquake and the sarin gas attack on the Tokyo subway, at a time of economic recession (2011a, 418, b, 86–8).[13] For Uno, the robot anime *Neon Genesis Evangelion* (1995) captures the tendency toward *hikikomori* (the Japanese phenomenon of social reclusiveness) of young boys in Japanese society who feel helpless and apathetic (2011b, 86). Its plot describes Shinji Ikari, a teenage boy protagonist, who is unlike the other boy characters in earlier fighting robot anime, like *Mobile Suit Gundam*, as he refuses to ride on a fighting robot. Indeed, Shinji's doubt about his organisation's 'righteous missions to defend the earth and humanity' and his existential quest have been discussed in terms of Japan's position in the 1990s as 'a nation that had recovered from the trauma of war only to find itself incapable of creating its own future' (Nukada and Murakami 2005, 88). This mentality of narcissism, hopelessness, and introversion shaped manga, anime, light novels, and games in the late 1990s and in the early 2000s, establishing a genre called *sekai kei*. In *sekai kei* works, male protagonists are extremely self-conscious and only concerned with their small, insular worlds and their romances with female superhuman or cyborg fighters who fight for humanity.[14] *Sekai kei* works provided the new generation of *otaku* with a way to feel close to virtual characters such as Hatsune Miku, creating their own small worlds of meaning on their computers. As examples of other cultural practices that facilitate the potent mix of Pygmalion fantasy, sexism, and insularity that forms the backdrop to Hatsune Miku's popularity with male fans, I will discuss the female idol (*aidoru*), a prevalent popular-culture form, in relation to the notion of *kawaii* (cuteness).

KAWAII AND FEMALE IDOLS

It is interesting to note that Hiroyuki Ito has not wanted Hatsune Miku to appeal only to the *otaku* market (T. Kubo 2011; Umezawa and Nakamura 2012). When designing the image of Hatsune Miku, Ito states he wanted a less sexualised image and chose an illustrator who could design a healthy-looking image of a young girl, using motifs that suggest 'technology', even though he wanted the voice of Hatsune Miku to be a 'Lolita voice', as I mentioned earlier (T. Kubo 2011). Ito clearly wants Hatsune Miku to participate in the mainstream cultural economy. In this regard, the clean image of Hatsune Miku recalls the carefully managed presentation of young celebrities in Japan that are called 'idols'. I will first examine Japanese idol culture in relation to the notion of *kawaii* and, second, the history of the virtual idol, which more closely resembles the situation of Hatsune Miku.

The concept of the idol in contemporary Japanese popular culture, in general, refers to men and women celebrities in their teens and early twenties—whether they are singers, musicians, actors, models, TV personalities, voice actors, or those who combine these professions—who become popular in the mass-media for a short period of time. There are male idols, but female idols are much more visible and popular. For the purposes of this chapter, I will discuss the general features of female idols in relation to the notion of *kawaii* (cuteness) and the apparatus of male visual consumption.[15]

Japanese female idols must project cuteness and innocence. In fact, a female idol becomes a sign for cuteness and innocence. This particular sign for femininity embodies a sense of vulnerability, naïveté, meekness, dependency, or ignorance, as femininity is defined within Japanese culture. Indeed, cultural anthropologist Hiroshi Aoyagi reports that an interviewee explains that his worshipping of idols aims to preserve 'the traditional adolescent femaleness that was becoming lost to the increased influence of American-style sexual liberation' (2005, 218). Female idols' vocal features, such as a high-pitched or nasal voice, or the use of girly expressions as understood within Japanese language conventions, are signs to be understood as reflecting this kind of adolescent femininity (Miller 2004, 153). There are male *otaku* fans who become excited by the so-called Lolita voice. The terms *koe ota* (voice *otaku*) or *seiyū otaku* (voice actor *otaku*) are used to describe the male *otaku* fans who become obsessed with voice actors. These visual and aural signs demarcate what is valued as an idealised image of cute femininity in contemporary Japan.

Kawaii (meaning cute, pretty, sweet, lovely, or adorable) is a Japanese adjective that is often used to praise female idols.[16] According to cultural theorist Inuhiko Yomota, as this concept underlies Japanese traditional aesthetics from the medieval period, *kawaii* is understood to be superficial, transient, profane, imperfect, or immature, as opposed to '*utsukusii*' (beautiful), which relates to sacredness, perfection, or eternity (2006, 36 and 76). While the latter term can indicate coldness or distance, the former term, emanated from one's bodily expressions, conveys a sense of physical closeness and familiarity that invites the desire to dominate or control a *kawaii* person who is regarded as inferior or lower (Yomota 2006, 76). Anthropologist Brain McVeigh similarly articulates that *kawaii* conveys 'power relations and power plays, effectively combining weakness, submissiveness, and humility with influence, domination, and control' (1996, 292).[17] Yomota also suggests that *kawaii*, in fact, hides a sense of the grotesque or abject, or a sense of otherness, whether it is used to describe a stuffed toy, a baby, or a young girl (2006, 111).[18] Importantly, there is a clear difference between the depiction of a stereotypical *kawaii* character in male *lolicon* manga and in manga for a female readership: while a *kawaii* character in male manga means a heroic, young girl with large breasts who is physically mature yet mentally still a child and who is submissive to men's orders, including their sexual advances, a typical *kawaii* girl character in female manga is portrayed in romantic terms, looks adolescent, and is less sexualised (Yomota 2006, 156–7). Both, however, show the young girl character as emotionally childlike and available for seduction (though the female manga renders the male as androgynous).

The sexual politics of *kawaii* in relation to the female idol become evident when considering the idols' supposed asexuality. Female idols need to be seen as inexperienced romantically and sexually, representing the idea of a virgin girl who is untouched and pure—the contemporary *shōjo*, once again. The revelation of an idol's relationship would be seen as scandalous and would damage her image. In February 2013, Minami Minegishi, a member of the popular idol group AKB48, appeared on television with a shaved head as public penance for sleeping with a man (News Limited 2013). There is also a slang term, '*koe buta*' (voice pig), to describe the fanatical male *otaku* fans of voice actors who are also regarded as idols; the *koe buta* want their voice-actor idols to remain virginal. Contrary to this idea of the female idol as virgin, there have also been idol groups with more adult themes, promoting the eroticism and sexiness of slightly older female performers.[19] According to Hiroshi Aoyagi, *otaku* worship of any of these female idols is meant be romantic rather than sexual (2005, 220). Cultural anthropologist Patrick W. Galbraith also reports a similar sentiment expressed by his interviewee, who claims that he is able to dissociate the soft-porn photographic or video image of his idols from the actual bodies of the idols, who therefore remain unspoiled (2012, 196–7). The double standards for the maintenance of *kawaii* and the virginal status of the female idol, and the image of the older idol as sexualised, support an impossible duality that leads men to desire unattainable and paradoxically virginal/sexual 'virtual' women (C. Ueno 2010, 19–20).

The unreachable and impossible female 'idol' in Japanese popular culture, a product for male consumption, results from a complex history of development and innovation within the entertainment industry. The term was initially used to refer to mainstream popular female singers who were admired by fans in the 1960s and 1970s. In the 1980s and 1990s, the focus of the term 'idol' shifted from talented individual singers to young female TV personalities within popular TV variety shows. These programs presented the process of turning 'the girl next door', often high-school girls, into idols, highlighting the mechanism of 'manufacturing' idols on TV (Endo 2013, 117). Fans increasingly accepted that a female idol is created and understood as a 'product' to be consumed and that the idol performs *kawaii* femininity through an obvious and schooled coquetry. In other words, according to Galbraith, '[m]en (especially *otaku*) were attracted to the idol as girl (*shōjo*), herself already a fiction of the consumer and of consumable pleasure' (2012, 195). Importantly, an idol must not be autonomous or create her own image, so, in effect, the male producer 'makes her an interchangeable and disposable image commodity' (Galbraith 2012, 193–4). However, according to Takashi Katsuki, writer on idol culture, part of an idol's appeal to male fans in terms of the idol industry and its known mechanics is that they can simultaneously see that an idol is controlled and manipulated as a 'puppet', and that there is an ordinary girl with her own subjectivity, working hard to be an idol, though she cannot embody the role perfectly (2014, 78). Her humility and imperfect assumption of the idol role contribute to her cuteness.

The popularity of Hatsune Miku among male *otaku* parallels their interest in female idol culture in terms of the aspects of control, interactive spectacle, and consumption. Given the history of the female idol in Japan, it is clear that Hatsune Miku, who is forever 16 years old, might be a virtual idol for the *otaku*, an idol on the Net. However, Hatsune Miku is not the first virtual idol in the history of the Japanese entertainment industry.

The first virtual idols appeared in the 1990s. Hikaru Ijuin, the producer of a popular radio program, All Night Nippon, broadcast by the Nippon Broadcasting System, presented an unidentified female singer/idol, Haga Yui (playing with words, meaning 'irritation'). The idea for Haga Yui was conceived in 1989, and 'she' performed in 1990 for a year. The publicly provided information about the singer was fabricated—for example, that the singer was a 15-year-old girl with a ponytail, and her height was 158 cm and weight 44 kg. While Haga Yui performed on the radio program and a CD was created, 'her' face was never shown to the public (Sawayaka 2008, 187). It was a 'tongue-in-cheek' project to prove that an idol can indeed be manufactured. Later, Yoshitaka Hori, the president of HoriPro, one of Japan's major production companies, took the virtual idol concept one step further and, in 1996, produced Date Kyoko, a singer represented by a 3D computer graphic model. Unlike the successful Haga Yui, for which the listeners and Ijuin had collaboratively shaped the image of the idol, Date Kyoko was regarded as unsuccessful due to its clumsy 3D image and its overdetermined, top-down conception. There was no space for fans to participate in the creation of Date Kyoko (Takahashi 2011, 142). Similarly, manga artist Ken'ichi Kutsugi's creation of the female CG model Terai Yuki appeared in the media in the late 1990s, but was not successful for similar reasons.

Outside the entertainment industry, Masataka Goto, a music technology research scientist and music software developer, and a leader of the media interaction group at AIST (the National Institute of Advanced Industrial Science and Technology), created a singing android in the context of scientific research. Goto is remembered for the development of music interface software: VocalListener and VocaWatcher. These technologies were installed onto a robot platform belonging to AIST, HRP 4C 'Miim' (where HRP 4C means Humanoid Robotics Project, 4th Cyborg), a female android. The demonstration performance of Miim, which sang and danced with human performers, was presented in 2010 at CEATEC (Combined Exhibition of Advanced Technologies). Through the synchronisation of singing voice, the sound of inhalations and exhalations, and facial expressions, that is, the movement of eyes, mouth, and the head, Goto aimed to replicate natural and realistic singing. Miim was not popular among the *otaku* as it was too literal.

What would determine the success for a virtual idol in male *otaku* subculture? For Daniel Black, theorist on the body and technology, '[t]he virtual [female] idol is a digital translation of a woman's body', and her success depends on what is gained or lost in terms of feminine attributes, and whether or not 'new attributes particular to her digital format' are present (2012, 218). The most important aspect of the virtual idol's digital aspects would be 'an availability for manipulation and modification, appropriation and control' (Black 2012, 219). The virtual idol Fujisaki Shiori, a cartoon character, was the main heroine of *Tokimeki Memorial*, a popular dating video game by Konami that was first released in 1994. As the game became more popular, Konami started to promote Fujisaki Shiori as a virtual idol by, for example, suppressing the name of voice actor Mami Kingetsu when producing a CD with Fujisaki Shiori as the singer. Fujisaki Shiori (voiced by Mami Kingetsu) participated in promotional events and radio shows. A decade later, in 2006, Namco created a popular game called *The Idolmaster*, in which a player produces his or her own female idol singer within the game. While Fujisaki Shiori presents an early example of a cross-platform product, Konami did not allow secondary creations, nor did Namco, so fans could not manipulate the character to create their own articulations of it. In contrast to these two, Hatsune Miku has succeeded because Ito's awareness of the importance of fan-based activities for the development of the Character Vocal Series allowed, even encouraged, appropriation of Hatsune Miku. Secondary creations of Hatsune Miku synergistically mediate fans' communication and participation.

The voice of Hatsune Miku is the primary vehicle for fan engagement, though fans are also keenly interested in its visual image, which they can appropriate and modify. The digitised and recomposed Lolita voice of Fujita in Vocaloid 2 is no longer a human voice, yet the original force of Fujita's human voice can be sensed, an affect that facilitates users to manipulate it (Ishida 2008, 91). For critic Sawayaka, the image designed by KEI and the 'personal' details offered about Hatsune Miku are carefully designed and honed to present a minimum of information, a 'clean iconography' that invites users' subjective inputs through secondary creations (2008, 184). Indeed, Ito sees Hatsune Miku as a formless medium to connect people, just like Bunraku dolls that allow space for the audience's projection of meaning (quoted in Shiba 2014, 259). Sawayaka regards that Hatsune Miku functions as a '*habu* (hub)' for fans' creativity (2014, 77).

The sexualised consumption of Hatsune Miku by male *otaku* embodies the most recent iteration of the modern vision of the artificial woman in the history of Western automata and their imagining in art and literature. These ideas, of creating, controlling, and possessing an ideal woman via techno-logical means, were imported to Japan and have contributed to popular-cultural expression. The fact that the Japanese term '*chōkyō*' (the training of an animal) is often used to describe the compositions and manipulations of Hatsune Miku via the Vocaloid engine suggests the persistence of the 'Pygmalion complex' (Masuda 2008, 38).

CIRCUITS OF SIGNIFICATION: *OTAKU* SEXUALITY

And what of the *otaku*'s own understanding of his engagement with cartoon characters, his points of identification? Japanese studies and Asian
studies scholar Setsu Shigematsu suggests that a male reader of *lolicon*
manga may switch between roles of male perpetrator and girl victim,
noticing 'the eroticization of the girls *and* being eroticized; vicariously
experiencing being the attacker by seeing the reactions of the cute girls
and imagining oneself to be the attacked/ stimulated young girl' (1999,
137, original emphasis).[20] The male *otaku* reader may also imagine himself
in the position of the onlooker, part of a triangulated gaze: 'the onlooker
and the attacker; or the little girl *and* the onlooker' (Shigematsu 1999,
137, original emphasis). So, in these various configurations, a male *otaku*
reader identifies himself with the girl character who gets attacked, from a
masochistic perspective (Akagi 1992, 232). At the same time, his perspective can quickly change to that of the attacker or onlooker, exhibiting
'sadism' through his 'scopophilic drive' (Shigematsu 1999, 134).

Crucially, the *otaku*'s highly coded anthropomorphism concerning
images means that these images are understood to be giving something
back in an imagined circuit of reciprocity. *Kyara moe* requires the presence
of the other. For example, Saito observes that while male *otaku* visually
consume images of young girls, these figures seem to look back at the *otaku*
with their large cartoon eyes, and the *otaku* feels he is being watched
(2011b, 220). The girl figure's imagined gaze at the *otaku* grants the
image an 'aliveness'. This reflexivity is an important aspect of *otaku*
in-group behaviour. That is to say that the *otaku*'s excitement concerning
these images of girls is moderated as a performance of enthusiasm, as a
'cool' gesture toward the other *otaku*, who might be more seriously
invested in these images. The performance is also an ironic and pleasurable
display for the non-*otaku* community, which might hold a negative view
of *otaku*. Hiroki Azuma discusses the *otaku*'s logic as the double operation
of a participant who follows the 'norm' of the *otaku* aesthetic and is also
the self-observing voyeur (2001, 100).

Paradoxically, although the figure of the *otaku* has become more acceptable to mainstream Japanese culture, images of the cute fighting girl are still
believed to be consumed in relation to Japanese cultural ideas of
paedophilia, sadism, masochism, or fetishism. Rather than rejecting these
labels, according to Saito, they provide the *otaku* with an 'alibi' for their
perceived perversity (Saito 2011a, 31). That is, the *otaku* are themselves

'performing' an *otaku kyara* who is regarded as a '*dame*' (useless or hope-less) person who cannot resist his 'pathetic' attraction to the cartoon images of young girls (Azuma 2007, 316).[21] For artist Takashi Murakami, they are defined by 'their relentless references to a humiliated self' (2005, 132). The label of *otaku* works synecdochically, as a separate entity, to represent and protect the person who wears it. At the same, functioning like the avatar in games, the *otaku* label allows a person declaring himself to be an *otaku* to experience something 'real' when consuming the young girl image in manga, anime, and games.

In the same vein, the *otaku*'s *kyara moe* creates yet another buffer. As I discussed above, stock figures, or *kyara*, are generated from a vocabulary of these characters, a 'database'. As Hiroki Azuma points out, the database is constituted on the basis of collective decisions made by the *otaku* commu-nity regarding what is real for them (2007, 60–1). This means that the *otaku* is following certain unspoken rules or codes regarding *kyara*. The message is that if you are an *otaku*, you *should* feel *moe* toward *kyara*. The duality of the relation of the performing self to the sign lies at the heart of the *otaku* gesture: an *otaku* is conscious of performing, being an *otaku*, and simulta-neously being an audience member within the *otaku* community—at the same time experiencing a real somatic responsiveness toward the object of affection. The *otaku* can present this performed self in an infinite loop of 'actor' and 'player', while enjoying being a spectator of his own actions.

Male fans of cartoon and fictional characters can hide their true selves in the performance of *otaku* sociality. Behind the buffer of in-group *otaku* performance, these imaginary figures, including the machine-woman that Hatsune Miku represents, are manipulated, controlled, consumed, and possessed—and imagined as responsive, wanting this kind of attention, in fact needing it to exist. The net-based, distributed practice of Hatsune Miku—as a figure of multiplicity, ubiquitousness, and, necessarily, empti-ness—facilitates not only the fantasy of the needy young girl requiring attention but also the homosocial bonding of male *otaku* at impassioned gatherings on the Internet sites and through the spectacles of online con-certs.[22]

Ironically, the popularity of Hatsune Miku has moved this character from its subcultural origins into the mainstream. This does not mean that the *hentai* interest in Hatsune Miku, the shadow region of Japanese popular culture, has disappeared. These practises can be understood in terms of the entertainments of '*akusho*', the marginal domain outside a society's sanctioned zones.

Moving on from Hatsune Miku as a virtual locus of inter-subjective and inter-corporeal engagement, the next chapter discusses the relationship between 'puppet and puppeteer' and the related identification of hobbyists with their toy robots, in the contexts of the robot competition.

NOTES

1. Hatsune Miku concerts have been held in major cities including Los Angeles (2011), Singapore (2012), Hong Kong (2012), Taipei (2012), Jakarta (2014), New York (2014), and Shanghai (2015). Composers such as Isao Tomita and Keiichiro Shibuya held concerts using the Hatsune Miku character for a high-art audience in 2012 and in 2013, respectively.

2. There are several *otaku* generations. The first generation of *otaku* was born around 1960. Each generation has different preferences with regard to their hobby activities (Azuma 2001, 13). Belonging to the oldest generation, Toshio Okada (who was in his mid-forties in 2005) defines *otaku* as an extreme collector or hobbyist who is 'social rejected', stating that he does not understand the phenomenon of *moe* amongst people under 35 years of age (Okada, Morikawa, and Murakami 2005, 173 and 177, original English).

3. The term 'subculture' in Japan is not understood as meaning 'counterculture' or 'alternative culture', that is, possessing a political edge. According to Sawaragi, this English term is often used in Japan in an abbreviated form as '*sabukaru*' that can be a pejorative term within the *otaku* lexicon, 'denoting [the older generations of] those followers of rock music and fashion imported from the West' (2005, 206, original English). Sawaragi nevertheless uses 'subculture' as 'a synonym for *otaku* culture, outside the Western dichotomy of high vs. low culture' (2005, 206, original English). For the English-language readership of this book, I use 'subculture' to refer to *otaku* culture as a distinctive cultural grouping based on shared interests within and separate from larger, mainstream Japanese culture.

4. On the topic of the sex-bot, see Levy (2007). For Ishiguro, it is a taboo topic for roboticists, though he is aware that his research potentially intersects with it (2009, 165–6).

5. For the history of the Japanese comics market, see Kinsella (2000) and Tamagawa (2012).

6. Women have been a leading force in Japanese manga fan culture from the 1970s, and there is a female subculture called *yaoi* (Kinsella 2000, 113–18; Saito 2007; Galbraith 2011).

7. LGBT issues in Japan have been discussed in relation to modernity. Same-sex relationships were socially accepted and were termed '*nanshoku*' (male eroticism) in the pre-modern Edo period, and, only after European sexology was introduced in the Meiji period of modernisation in the late nineteenth century, '*nanshoku*' was discussed in relation to imported concepts such as 'perverse' or 'queer' (McLelland et al. 2007, 6–7).

8. The Japanese word *hentai* has been exported to the porn industry in English-speaking countries to refer to Japanese pornographic comics, animation, and games. Video game theorist Ian Bogost discusses the very violent, disturbing, and misogynistic nature of these Japanese porn products as a lesson in 'the various logics of perversion that stimulate other human beings' (Bogost 2011, 109).

9. Child pornography has been policed since the Law for Punishing Acts Related to Child Prostitution and Child Pornography and for Protecting Children was implemented in Japan in 1999. In 2010, the Tokyo metropolitan government expanded the scope of the Tokyo Metropolitan Ordinance Regarding the Healthy Development of Youth by making the publishing industry regulate publications that depict persons under 18 in sexual and violent contexts.

10. The 'Lolita complex' is usually understood in relation to the Hollywood film *Lolita* (1962), based on Vladimir Nabokov's novel of the same title (1955). According to writer Yasushi Takatsuki, Lewis Carroll's *Alice's Adventures in Wonderland* (1865) has actually been more popular within male *otaku* culture (2009, 34).

11. Other studies on Oshii's work include those of Bolton (2002), Orbaugh (2007), and Brown (2010).

12. *Lolicon* desire can also be found in Japanese literary classics, such as the story of 'Lady Murasaki' in the eleventh-century classic, *Tale of Genji*, or in the story of Naomi in *A Fool's Love* (1924), a modern classic novel by Jun'ichiro Tanizaki: these works aestheticise men's dreams of forbidden, predatory love toward a daughter or express a dream of raising a young girl with the intention of marrying her (C. Ueno 2010, 162–3). The culture of *lolicon* resonates with the masochistic and self-parodic nature of Tanizaki's novel, in which the male protagonist is overpowered by the young heroine. The novel is

discussed as a 'strategic cultural retreat' from the rapid modernisation of the 1920s (Najita 1989, 12). At the same time, the '*moga*' (modern girl) gained 'a measure of autonomy and financial independence' in the 1920s, which challenged the Japanese patriarchal tradition (Harootunian 2000a, 119).

13. The gas attack was perpetrated by the members of Aum Shinrikyo, a cult group, and killed more than ten people.

14. Typical *sekai kei* works include *Neon Genesis Evangelion* (1995), *Saikano: The Last Love Song on This Little Planet* (2000), *Iriya's Sky, Summer of the UFOs* (2001), and *Voices of a Distant Star* (2002).

15. For discussions of the male idol, see Nagaike (2012) and Glasspool (2012).

16. *Kawaii* style has been a noteworthy social and cultural phenomenon in Japan since the 1980s. Significantly, in 2009, the Ministry of Foreign Affairs appointed three female 'ambassadors of cute' for the promotion of Japan's contemporary popular culture.

17. *Kawaii* connects to the meaning of '*kawaisō*' (pitiable), which triggers empathetic responses in a hierarchical way, as in '*kawaigaru*' (to love, to fondle, to make a pet of, or to be rough with) (McVeigh 1996, 301).

18. The slang word '*kimokawa*', which is *kawaii* combined with '*kimochi warui*' (disgusting), meaning 'disgusting yet cute', expresses the ambivalence of *kawaii* (Yomota 2006, 89).

19. For example, the Ebisu Muscats (2008–2013, 2015–present) is a group that consists of '*gurabia*' idols, which are popular models for erotic photography magazines and adult video actors.

20. Takatsuki reports that his *otaku* creator interviewee acknowledges his desire to be a young girl (2009, 121).

21. While Toshio Okada acknowledges that the notion of *dame* is applicable to the younger generations of *otaku*, he rejects the term to describe *otaku* culture as a whole (Okada et al. 2005, 181).

22. I agree with Thomas LaMarre's reading of Saito's point that the homosociality of *otaku* culture as a 'heterosexual therapy for the computer age' maintains heterosexual normativity (2009, 257). However, Japanese heterosexual normativity may not be monolithic, and clearly can accommodate *otaku* activities situated between 'the *otaku*'s actual heterosexual "wholesomeness" and the polymorphous perversity of their fantasies' (Vincent 2011, xx).

REFERENCES

Akagi, A. (1992). Bishōjo shōkō gun, Rorikon toiu yokubō [The beautiful young girl syndrome, the desire called Lolicon]. *New Feminism Review, 3*, 230–234.

Aoyagi, H. (2005). *Islands of eight million smiles: Idol performance and symbolic production in contemporary Japan.* Cambridge, MA: Harvard University Asia Center.

Azuma, H. (2001). *Dōbutsuka suru posutomodan: Otaku kara mita nihonshakai* [Animalizing postmodern: Japanese society as seen from *otaku*]. Tokyo: Kōdansha.

Azuma, H. (2007). *Gē mu teki riarizumu no tanjō: Dōbutsuka suru posutomodan 2* [The birth of game realism: Animalising the Postmodern 2]. Tokyo: Kōdansha Gendai Shinsho.

Black, D. (2012). The virtual idol: Producing and consuming digital femininity. In P. W. Galbraith & J. G. Karlin (Eds.), *Idols and celebrity in Japanese media culture* (pp. 209–228). Houndmills/Basingstoke/Hampshire/New York: Palgrave Macmillan.

Bogost, I. (2011). *How to do things with videogames.* Minneapolis: University of Minnesota Press.

Bolton, C. A. (2002). From wooden cyborgs to celluloid souls: Mechanical bodies in anime and Japanese puppet theater. *Positions: East Asia Cultures Critique, 10*(3), 729–771.

Brown, S. T. (2010). *Tokyo cyberpunk: Posthumanism in Japanese visual culture.* New York: Palgrave Macmillan.

Crypton Future Media. (2016a). Hatsune Miku V4X. http://www.crypton.co.jp/mp/pages/prod/vocaloid/mikuv4x.jsp. Accessed on 10-10-2016.

Crypton Future Media. (2016b). Vocaloid 2, Hatsune Miku. http://www.crypton.co.jp/mp/pages/prod/vocaloid/cv01.jsp. Accessed on 04-15-2016.

Endo, T. (2013). *Sōsharuka suru ongaku: Chōshu kara asobi e* [Music facilitates sociability: From listening toward playing]. Tokyo: Seidosha.

Enomoto, A. (Ed.). (2009). *Otaku no kotoga Omosiroi hodo wakaru Hon* [This book easily allows you to understand what an *otaku* is]. Tokyo: Chūkei shuppan.

Fiske, J. (1992). The cultural economy of fandom. In L. A. Lewis (Ed.), *The Adoring audience fan culture and popular media* (pp. 30–49). London/New York: Routledge.

Galbraith, P. W. (2011). Fujoshi: Fantasy play and transgressive intimacy among 'Rotten Girls' in contemporary Japan. *Signs, 37*(1), 211–232. doi:10.1086/660182.

Galbraith, P. W. (2012). Idols: The image of desire in Japanese consumer capitalism. In P. W. Galbraith & J. G. Karlin (Eds.), *Idols and celebrity in Japanese media culture* (pp. 185–208). Houndmills/Basingstoke/Hampshire/New York: Palgrave Macmillan.

Galbraith, P. W. (2015). *Otaku* sexuality in Japan. In M. J. McLelland & V. Mackie (Eds.), *Routledge handbook of sexuality studies in East Asia* (pp. 205–217). London/New York: Routledge.

Glasspool, L. (2012). From boys next door to boys' love: Gender performance in Japanese male idol media. In P. W. Galbraith & J. G. Karlin (Eds.), *Idols and celebrity in Japanese media culture* (pp. 113–130). Houndmills/Basingstoke/ Hampshire/New York: Palgrave Macmillan.

Hamano, S. (2008). *Ā kitekuchā no seitai kei: Jōhō kankyō wa ikani sekkei sarete kitaka* [Ecology of architecture: How the information environment has been designed]. Tokyo: NTT Shuppan.

Harootunian, H. D. (2000a). *History's disquiet modernity, cultural practice, and the question of everyday life*. New York: Columbia University Press.

Huyssen, A. (1986). *After the great divide: Modernism, mass culture, postmodernism*. Bloomington: Indiana University Press.

Ishida, M. (2008). Naka no hito ni naru: Koe modoki ga kanōni shita mono [Becoming an insider: What imitation voice can allow]. *Yuriika, 40*(15), 88–94.

Ishiguro, H. (2009). *Robottoto wa nanika: Hito no kokoro wo utsusu kagami* [What is the robot?: A mirror reflecting the human soul]. Tokyo: Kōdansha Gendaishinsho.

Ito, G. (2005). *Tezuka izu deddo: Hirakareta manga hyōgenron e* [Tezuka is dead: Open expression in manga]. Tokyo: NTT Shuppan.

Ito, M. (2012). Introduction. In M. Ito, D. Okabe, & I. Tsuji (Eds.), *Fandom unbound: Otaku culture in a connected world* (pp. xi–xxxi). New Haven: Yale University Press.

Iwabuchi, K. (2002). 'Soft' nationalism and narcissism: Japanese popular culture goes global. *Asian Studies Review, 26*(4), 447–469. doi:10.1080/ 10357820208713357.

Katsuki, T. (2014). *Aidoru no yomikata* [A way to read idols]. Tokyo: Seikyūsha.

Kijima, Y. (2012). The fighting gamer otaku community: What are they 'fighting' about?. In M. Ito, D. Okabe, & I. Tsuji (Eds.), *Fandom unbound: Otaku culture in a connected world* (pp. 249–274). New Haven: Yale University Press.

Kinsella, S. (2000). *Adult manga: Culture and power in contemporary Japanese society*. Richmond/Surrey: Curzon.

Kubo, T. (2011). Special interview. *Inter-X-Cross Creative Center.* http://www. icc-jp.com/special/2011/01/001726.php. Accessed on 04-15-2013.

Kubo, A. (2015). *Robotto no jinruigaku: Nijūseiki no kikai to ningen* [Anthropology of the robot: The machine and the human in the 20th century]. Tokyo: Sekai Sisōsha.

LaMarre, T. (2009). *The anime machine: A media theory of animation*. Minneapolis: University of Minnesota Press.

Levy, D. N. L. (2007). *Love and sex with robots: The evolution of human-robot relations.* New York: HarperCollins.

Masuda, S. (2008). Hatsune miku kara tōku hanarete [Away from Hatsune Miku]. *Yuriika, 40*(15), 184–192.

McLelland, M., Suganuma, K., & Welker, J. (2007). Introduction: Re(claiming) Japan's queer past. In M. McLelland, K. Suganuma, & J. Welker (Eds.), *Queer voices from Japan: First person narratives from Japan's sexual minorities* (pp. 1–29). Lanham: Lexington Books.

McVeigh, B. (1996). Commodifying affection, authority and gender in the everyday objects of Japan. *Journal of Material Culture, 1*(3), 291–312. doi:10.1177/135918359600100302.

Miller, L. (2004). You are doing Burikko!: Censoring/scrutinizing artificers of cute femininity in Japanese. In S. Okamoto & J. S. Shibamoto Smith (Eds.), *Japanese language, gender, and ideology cultural models and real people* (pp. 148–165). New York: Oxford University Press.

Miller Frank, F. (1995). *The mechanical song: Women, voice, and the artificial in nineteenth-century French narrative*. Stanford: Stanford University Press.

Morikawa, K. (2011). Azuma Hideo wa ikanishite otaku bunka no so ni nattaka [How did Hideo Azuma become the founder of *otaku* culture]. In *Azuma Hideo: Bishōjo, SF, fujōri gyagu, soshite shissō, sōtokushū* [Beautiful girl, SF, nonsense gag, and disappearance: A special issue] (pp. 179–186). Kawade Yumemukku Bungei Bessatsu. Tokyo: Kawade Shobō Shinsha.

Murakami, T. (2005). Earth in my window. In T. Murakami (Ed.), *Little boy: The arts of Japan's exploding subculture* (pp. 99–149). New York/New Haven/London: Japan Society and Yale University Press.

Nagaike, K. (2012). Johnny's idols as icons: Female desires to fantasize and consume male idol images. In P. W. Galbraith & J. G. Karlin (Eds.), *Idols and celebrity in Japanese media culture* (pp. 97–112). Houndmills/Basingstoke/Hampshire/New York: Palgrave Macmillan.

Najita, T. (1989). On culture and technology in postmodernism and Japan. In M. Miyoshi & H. D. Harootunian (Eds.), *Postmodernism and Japan* (pp. 3–20). Durham: Duke University Press.

News Limited. (2013). Japan AKB48 Pop Idol Minami Minegishi Shaves Head in Penance for Spending Night with Man. *news.com.au*. http://www.news.com.au/entertainment/music/japan-akb48-pop-idol-minami-minegishi-shaves-head-in-penance-for-spending-night-with-man/story-e6frfn09-1226567125292. Accessed on 10-04-13.

Nimiya, K. (2009). Takarazuka to shōjo to moe [Takarazuka, Girls, and '*Moe*']. In Sekyūsha Henshūbu (Ed.), *Takarazuka toiu sōchi* [Takarazuka as an apparatus] (pp. 307–340). Tokyo: Sekyūsha.

Nukada, H., & Murakami, T. (2005). Little boy (plates and entries). In T. Murakami (Ed.), *Little boy: The arts of Japan's exploding subculture* (pp. 1–97). New York/New Haven/London: Japan Society and Yale University Press.

Okada, T., Morikawa, K., & Murakami, T. (2005). Otaku talk. In T. Murakami (Ed.), *Little boy: The arts of Japan's exploding subculture* (Reiko Tomii,

Trans.) (pp. 165–185). New York/New Haven/London: Japan Society and Yale University Press.

Orbaugh, S. (2007). Sex and the single cyborg: Japanese popular culture experiments in subjectivity. In C. Bolton, I. Csicsery-Ronay, & T. Tatsumi (Eds.), *Robot ghosts and wired dreams Japanese science fiction from origins to anime* (Christopher Bolton, Trans.). (pp. 222–249). Minneapolis: University of Minnesota Press.

Otsuka, E. (1997). *Shōjo minzokugaku: Seikimatsu no shinwa wo tsumugu, miko no matsuei* [Ethnology of the girl: Spinning a myth at the end of the century, descendants of the Japanese Shrine Maiden]. Tokyo: Kōbunsha.

Otsuka, E. (2001). Azuma Hideo: Otakunaru mono no kigen [Hideo Azuma: The origin of the *otaku*]. In E. Otsuka & G. Sasakibara (Eds.), *Kyōyō toshite no manga, anime* [Manga, anime as education] (pp. 85–108). Tokyo: Kōdansha Gendai Shinsho.

Otsuka, E. (2009). *Atomu no meidai: Tezuka Osamu to sengo manga no shudai* [An atom thesis: Osamu Tezuka and themes in Postwar Manga]. Tokyo: Kadokawa shoten.

Otsuka, E. (2012). *Monogatari shōhiron kai* [A theory of narrative consumption, revised edition]. Tokyo: Kadokawa Gurūpu Paburisshingu.

Otsuka, E., & Osawa, N. [Nobuaki] (2005). *Japanimeishon wa naze yabureruka* [Why 'Japanimation' will be defeated]. Tokyo: Kadokawa Shoten.

Robertson, J. (2010). Gendering humanoid robots: Robo-sexism in Japan. *Body & Society, 16*(2), 1–36. doi:10.1177/1357034X10364767.

Saito, T. (2007). Otaku sexuality. In C. Bolton, I. Csicsery-Ronay, & T. Tatsumi (Eds.), *Robot ghosts and wired dreams Japanese science fiction from origins to anime* (C. Bolton, Trans.) (pp. 222–249). Minneapolis: University of Minnesota Press.

Saito, T. (2011a). *Beautiful fighting girl* (J. K. Vincent & D. Lawson, Trans.). Minneapolis: The University of Minnesota Press.

Saito, T. (2011b). *Kyrakutā seishin bunseki: manga, bungaku, nihonjin* [*Character psychoanalysis: Manga, literature, the Japanese*]. Tokyo: Chikuma Shobō.

Sasakibara, G. (2004). *Bishōjo no gendaishi: Moe to kyarakutā* [A modern history of the beautiful girl: 'Moe' and character]. Tokyo: Kōdansha.

Sawaragi, N. (2005). On the battlefield of 'Superflat': Subculture and art in postwar Japan. In M. Takashi (Ed.), *Little boy: The arts of Japan's exploding subculture* (L. Hoaglund, Trans.) (pp. 187–207). New York/New Haven/London: Japan Society and Yale University Press.

Sawayaka. (2008). Kumiawasareru shōjo [A constructed girl]. *Yuriika, 40*(15), 184–192.

Sawayaka. (2014). *Jūnendai Bunkaron* [Cultural theory concerning 2010s]. Tokyo: Seikaisha Shinsho.

Shiba, T. (2014). *Hatsune miku wa naze sekai wo kaetanoka* [Why has hatsune Miku changed the world]. Tokyo: Ōta Shuppan.

Shigematsu, S. (1999). Dimensions of desire: Sex, fantasy, and fetish in Japanese comics. In J. A. Lent (Ed.), *Themes and issues in Asian cartooning: Cute, cheap, mad, and sexy* (pp. 127–163). Bowling Green: Bowling Green State University Popular Press.

Stevens, C. S. (2010). You are what you buy: Postmodern consumption and fandom of Japanese popular culture. *Japanese Studies, 30*(2), 199–214. doi:10.1080/10371397.2010.497578.

Takahashi, N. (2011). *Bōkaroido genshō: Sinsē ki kontentsu sangyō no mirai moderu* [The phenomenon of Vocaloid: The future model for content business in the new century]. Tokyo: PHP Kenkyūjo.

Takatsuki, Y. (2009). *Rorikon: Nihon no shōjo shikōsha tachi to sono sekai* [Lolicon: Japan's Shōjo Lovers and Their World]. Tokyo: Basilico.

Tamagawa, H. (2012). Comic market as space for self-expression in otaku culture. In M. Itō, D. Okabe, & I. Tsuji (Eds.), *Fandom unbound: Otaku culture in a connected world* (pp. 107–132). New Haven: Yale University Press.

Ueno, C. (2010). *Onnagirai: Nippon no misoginī* [Woman Hater: Japanese Misogyny]. Tokyo: Kinokuniya Shoten.

Umezawa, M., & Nakamura, Y. (2012). Hatsune Miku genshō ga hiraku kyōkanryoku no shinsekai [The New World of Empathy Through the Hatsune Miku Phenomenon]. Interview with Hiroyuki Ito, Yasuharu Ishikawa, and Toshiyuki Inoko. *Toyo Keizai On-Line.* http://toyokeizai.net/articles/-/11315. Accessed on 15-04-2013.

Uno, T. (2011a). *Ritoru pīpuru no jidai* [The age of little people]. Tokyo: Gentōsha.

Uno, T. (2011b). *Zeronendai no sōzōryoku* [The 'Noughties' imagination]. Tokyo: Hayakawa Shobō.

Vincent, J. K. (2011). Translator's introduction, making it real: Fiction, desire, and the queerness of the beautiful fighting girl. In S. Tamaki (Ed.), *Beautiful fighting girl* (J. K. Vincent & D. Lawson, Trans.) (pp. ix–xxv). Minneapolis: University of Minnesota Press.

Yomota, I. (2006). *Kawaii ron* [A theory of cute]. Tokyo: Chikuma Shobō.

Yoshida, M. (2004). *Nijigen bishōjo ron* [A theory of the two-dimensional beautiful girl]. Tokyo: Futami Shobō.

Yoshimoto, T. (2009). *Otaku no kigen* [The origin of the *otaku*]. Tokyo: NTT shuppan.

CHAPTER 7

Competition Robots: Empathy and Identification

Robot hobbyist contests participate in a kind of grass-roots robot-making in Japan. Hobbyist competitions are often included in the promotion of next-generation robots, and they represent a local version of larger forums in which the close relationship between humans and robots is a pervasive theme. These robot competitions are organised events for purpose-built hobby robots where they sprint, play soccer, or fight each other. There are numerous robot competition events every year for junior high-school students, high-school students, university students, and adults. Out of these, this chapter looks into two types of competitions that present anthropomorphic robots: ROBO-ONE, a popular hobbyist competition since 2002 that features robot fighting, and Bacarobo, a staged comedy contest for homemade robots that are designed to entertain the audience with laughter rather than to fight each other.[1] I discuss Bacarobo events held in 2007 and 2008.

Outside the usual framework of function-focused robot displays, these events present situations in which hobby robots with individual characteristics entertain their audiences through games and comedy, respectively. I will discuss how the specific *mises en scène* of these events help to establish the 'presence' and purpose of hobby robots: for the Japanese robot enthusiasts at these events, the competition robots become charged figures. '[O]bjects are not [simply] means, but rather mediators', as sociologist Bruno Latour suggests (1996, 240). These contest robots are of diminutive dimensions and can look like toys, but they powerfully negotiate the bonds of temporary communities formed in the course of these competitions. As I will discuss,

© The Author(s) 2017
Y. Sone, *Japanese Robot Culture*,
DOI 10.1057/978-1-137-52527-7_7

they are also reflective of the Japanese view of the relation of the human to the non-human world. The competition context can turn a machinic object into a figure of deep resonance for participants at these events.

In the first half of this chapter, I examine the generative relationship between hobby robot creators and their robot agents at ROBO-ONE, where robot 'athletes' that are controlled by their creators/contestants at the side of the stage do battle with each other. ROBO-ONE's one-to-one combat style facilitates an anthropomorphising of the participating robots, which are built to have unique appearances. ROBO-ONE raises pressing questions, not least of all 'who is performing?' (is it the human controller, or the machine?) and 'what is performed?' (i.e., what social and cultural concerns are played out in this performance/fight/game?). In the chapter's second half, I discuss Bacarobo, which involves more complexly reflexive relations between robots and human participants through comedy. I will explain how Bacarobo can generate a temporary communal atmosphere reminiscent of the ancient *matsuri*.

The *mises en scène* of ROBO-ONE and Bacarobo grant the performing robots a special, limited agency. I consider how these robot competitions build communities of robot supporters that at the same time demarcate, as occurs in many other domains of Japanese society, what is 'inside' and what is 'outside'. I will first discuss the history of the robot contest in Japan.

ROBOT CONTESTS

Robot competitions have been an important part of the Japanese robotics landscape since the 1980s.[2] They are seen as local examples of *monozukuri* (the art of making things), a phrase often used by Japanese manufacturers and the government to promote Japanese technology and culture. The best-known competition is Robocon, the shortened version of the English phrase 'robot competition', organised for college and university students.[3] In Robocon, homemade, remote-controlled robots of non-anthropomorphic appearance compete by carrying out a specified task in a set time. The original event was initiated by Masahiro Mori in 1988. Mori felt there was an educational need to stimulate his uninspired university students, emphasising creativity and enjoyment in the process of making (2014, 272). Robocon has been broadcast annually since 1991 by the national broadcasting organisation NHK (Japan Broadcasting Corporation), and it has been developed into regular, major national and international robot competitions for students from junior high school to university. RoboCup, on the other hand, is a much larger international annual event for robot soccer, inaugurated in 1997. It attracts thousands of participants from more than 40 countries, involving university researchers and students. RoboCup is much more complex in terms of the technologies of the robots involved and is intended to provide opportunities for academic research in artificial intelligence and robotics.

The most notable and prominent robot competition events have been conceived and organised by roboticists, with the aim of developing robotics research and education. ROBO-ONE, on the other hand, is supported by hobby robot companies, that is, those that manufacture robot kits and sell hobby supplies. Importantly, compared with the more seriously inclined robot competitions, which also use forms of sports in their contests, ROBO-ONE is much more concerned with the event's playful and entertaining aspects (Hotta 2008, 165).[4] Similarly non-research-oriented, Bacarobo was originally conceived by an artist, and it is supported by an entertainment production company. The individuality of the participating anthropomorphic robots as well as aspects of surprise and drama in the staging of its contests are all important features of ROBO-ONE and Bacarobo.

ROBO-ONE

ROBO-ONE runs biannual combat competitions of remote-controlled model humanoids in Japan. At ROBO-ONE, two small-sized, metallic robots of roughly human-like appearance, typically 30 to 50 centimetres tall, do battle in a fighting arena, a ring of around three metres in diameter.[5] While shuffling forwards, sideways, or backwards, these bipedal, remote-controlled robots try to take their opponents down or force them outside the ring by punching, pushing, or shoving.[6] Most of these bipedal machines are made in whole or in part from hobby robot kits available on the retail market. These robots can be programmed to walk and run, as well as to perform gymnastics, dance routines, or combat-related movements. They can be dressed up as mechanical warriors, doll- or animal-like characters, or as fantasy figures (See Fig. 7.1).

ROBO-ONE is foremost a contest, in that it is governed by rules and structures; however, while some contestants no doubt compete for prize money, ROBO-ONE is also a festival for hobby robot enthusiasts. Rather than taking the winning and losing of matches too seriously, the organisation emphasises and encourages entertainment and celebrates the dexterity of participating robotic machines (Azusa 2009). While the majority of the

Fig. 7.1   ROBO-ONE, the 18th staging of this event in Niigata, Japan, 2010 © Biped Robot Association (Courtesy of Biped Robot Association)

contestants are male adults who often have engineering backgrounds, their wives and children almost always participate. It is promoted as a family-friendly event that strives to generate a celebratory atmosphere for contestants, organisers, and the audience.

Even though its hobby robots are pitted against each other in staged fights, neither the literature around the competition and its video documentation nor the atmosphere of the event itself suggests any sense of aggression. Its atmosphere is quite different from similar events in the USA. For example, the American competition RoboGames (formerly ROBOlympics, which began in 2004) presents aggressive 'caged' combat of remote-controlled robotic vehicles. *Robot Combat League*, a TV contest for 'staged', violent fights involving human-sized humanoids, with fire belching and sparks flying, aired in the USA in 2013.[7] Philosopher Jean Baudrillard's understanding of robot-human relations, which is that 'the anticipated disintegration of the robot produces a strange satisfaction' (2005, 132), accurately describes the tone of these robot shows. This kind of aggression toward robots is totally absent in ROBO-ONE. How can we explain this difference between ROBO-ONE and its Western counterparts? How does ROBO-ONE articulate a particularly Japanese response to the robot?

At first glance, the captivating immediacy of the competition aspect of ROBO-ONE appears to be the main reason for the bonds among the participants, officials, and audience. As well, though, ROBO-ONE creates common ground for its Japanese contestants *in relation to* their robots; this is a particularly reflexive relationship. And accompanying these modes of identification is a shared sense of awe and reverence toward machinic objects, even if its robot-on-robot fights may seem comical. This seemingly contradictory understanding becomes less opaque when we view the robots from the Buddhist perspective of the self and Japanese cosmology, which allows for a reflexive anthropomorphism in relation to the adored robots. In the Japanese context, there can be close inter-categorical connections that blur the distinction between organic beings and inanimate objects. In ROBO-ONE the three-dimensional, machinic objects can occupy an inter-mediary state between human and non-human being—a state that I have indicated earlier in this book as characterising the Japanese reception of the robot—which is activated in the course of the game.

Before discussing this notion of inter-categorical connection, I will, first, discuss anthropologist Victor Turner's concept of 'communitas' in order to articulate the transformative nature of games (1982).

THE PUPPET, THE PUPPETEER, AND THE SPECTATOR

While ROBO-ONE allows autonomous robots, many contestants use remote-controlled robots, as well as robots that use both systems. A doll-like object, carried by its owner, is made to stand at the edge of the fighting ring. As soon as the power is switched on, it comes to life. The robot is a kind of puppet, a stand-in for a visible but sidelined contestant at the edge of the fighting stage. The relationship between a human contestant and his or her remote-controlled robot can also be understood as related to telepresence, a concept which usually describes a person experiencing two distant places at once through a digital communication network. The coupling of robotics with teleconference technologies (as for Ishiguro's Geminoids, as discussed in Chap. 4, or in current medical science applications, for example) is a form of telepresence in which there is an object at the other end of the distant connection. The puppet-puppeteer relation highlights an awareness of control and ownership concerning the actions of a machinic object at a distance, combined with an awareness of one's own engrossment in an artificial environment. The opposing contestants at each end of the stage meet through their fighting avatars. While the contestants engage in robot matches, they also each observe the other contestant's robot performing. The contestants are simultaneously actors and spectators who appreciate the skill with which each other's robots were constructed and their opponent's dexterity in operating them.

Non-contestant observers (the officials and the spectators) become captivated by the infectious power of the game. Once a spectator starts watching the contest, he or she may favour a particular team of robot-human contestants. Whether taking the proceedings seriously or not, viewers get hooked into the contest by supporting a particular side in the same way as supporters barrack for their favourite teams in other sports. A spectator may even develop an empathetic feeling toward a robot-human pair.

This kind of affective spectatorship can be better understood in relation to Victor Turner's notion of 'communitas', referring to the temporary affiliation of a group of people who come together as a function of the unifying power of a situation (Turner 1988, 84). It is important to note here that a common use of Turner's communitas has been critiqued as celebrating positive, liberating experiences, dismissing dystopian or fascistic forms of spontaneous community (Maxwell 2008, 60). I use Turner to examine the connective mechanisms of ROBO-ONE but do not see ROBO-ONE in terms of salvation or liberation. ROBO-ONE brings its participants together as a symbolic community. However, the sense of communitas that occurs in ROBO-ONE also includes a non-human element, the competing robots, complicating ideas of spontaneous, shared experience.

Leisure is understood by Turner as '*freedom to* enter, even to generate new symbolic worlds of entertainment, sports, games diversions of all kinds' and to 'transcend social structural limitations' (Turner 1982, 37, original emphasis). As I suggested earlier, ROBO-ONE is clearly marked as entertainment even though it is also a contest that requires the serious efforts of the contestants. In such leisure activities, Turner adds, effort 'is chosen voluntarily, in the expectation of an enjoyment that is disinterested, unmotivated by gain, and [this investment] has no utilitarian or ideological purpose' (1982, 37). ROBO-ONE is thus an exemplary form of serious play.

The capacity of leisure to transcend social limitations becomes clear when Turner discusses the notion of 'communitas' in relation to the concept of 'flow'. Flow is understood as 'a state in which action follows action according to an internal logic, with no apparent need for conscious intervention on [the participants'] part' (Turner 1988, 54). Sports, games, art, and hobbies are thought to produce engrossed experiences for those who participate in these leisure and aesthetic activities, generating a sense of 'shared flow', or communitas (Turner 1988, 133).

The assimilative power of games can help us understand how a shared experience or a 'community' might include objects or things. Because of the rule-binding nature of games, a participant is made to focus on the game's activities, forgetting social realities that exist outside the game. The unifying power of flow temporarily generates a collective feeling beyond the boundaries of the self. The intense focus induced through the flow experience of a game unites all participants, gamers, and viewers, no matter what their ontological status may be: 'all men, *even all things,* are felt to be one, subjectively, in the flow experience' (Turner 1982, 57, my emphasis). The flow experiences of the puppet, the puppeteer, and the audience unite the humans and the robots in ROBO-ONE.

I will now focus on the relation at the centre of this matrix, that between ROBO-ONE contestants and their participating robots.

Robots as Mediators

The film documentary *Astroboy in Roboland* (2008), on robot culture in Japan, includes an interview with a Japanese family of three (parents and their daughter) who are regular participants of ROBO-ONE-style fighting robot competitions in Japan. The father states, 'The whole family enjoys the robots. The robots benefit our family life. These robots help us strengthen the bonds among the family members' (Caro 2008, English subtitles from the documentary). The view that hobby robots provide the means to improve family relationships, a process often attributed to other sports or any shared hobby, is reasonable and unremarkable; but the mother's comments reveal a less obvious dimension to the benefits conferred by family involvement in robot competitions. She discusses her maternal feelings for the robots:

> I feel attached to the robots. And I make their costumes. They speak, too. To see them move, I feel they're like my children. We put our hearts and souls into our robots. So they're not pieces of machine or aluminium objects. They are our children. They're as sweet as our daughter used to be before she became brash. (Caro 2008, English subtitles from the documentary)

Here, the mother seems to echo Baudrillard's assessment of the hierarchy between humans and machines in Western thought, in that there is:

an obedient functionality embodied (so to speak) in an object which resembles me, an object to which the world is subject yet which is simultaneously subject to my will. In this way a threatening part of myself has been exorcized and turned into a sort of all-powerful slave, cast in my image, which I can use for purpose of self-aggrandizement. (2005, 130)

One could also say that the mother's attitude to her hobby robots is a reflection of one's attachment to personally owned '"utilitarian" items such as car, motorcycles and boats' (Borenstein and Pearson 2012, 258). Anthropomorphising and individualising personally owned objects is not just a Japanese phenomenon. There are certainly also Western adults who anthropomorphise robots. For example, writer Lydia Pyne indicates her anthropomorphic view on her family's motorised floor mop, named Isaac, which is regarded as a pet or a child (2014). But there is also a particularly Japanese sensibility inherent in the apparent equality between the humans in the Japanese family and its robot members. There is more to the mother's comment than self-aggrandisement or the exercising of an unequal power relationship. Pyne's anthropomorphism, on the other hand, is a 'convenient' intellectual device to accommodate the machine's behaviour, its limited mobility and cleaning capability, providing 'a comfortable enough distance to make and re-make the robot as we best see fit' (2014). The Japanese mother's articulations make more sense when we examine how Japanese Buddhism and Japanese myth shape attitudes to the inanimate in Japanese society.

While there is no way of knowing if the mother who comments on robots as if they were family members was influenced by Japanese traditional thinking as such, these beliefs are often cited in to the literature on Japanese robots. Although my aim is not to essentialise the Japanese 'affinity' for robots or participate in *nihonjinron* narratives, which I have critiqued in earlier chapters, it is useful to recognise Japan's animist past, which inflects contemporary life as one contributor to Japanese attitudes and in relation to the contexts that shape the Japanese reception of robots. Animism is a kind of cultural 'affordance', a term I used in Chap. 4, per James J. Gibson and Donald Norman, to explain the importance of sociocultural knowledge and past experiences. While Japanese roboticists often refer to affordance in universalised and non-culturally specific ways, I use it here in culturally specific and functional terms to refer to a reading of available cultural and social information in the Japanese context to determine one's next move; in the case of robots, it is a matter of adapting cultural knowledge in a social situation to know how one might accommodate a robot. Clearly distinct from its Western version, there is no clear division between humans and non-humans in Japanese anthropomorphism. Animist beliefs can facilitate bonds with inanimate objects, such as robots, especially through the transformative power of game and sports. To explain this further in terms of the robot competition, I now turn to Masahiro Mori's Buddhist philosophy on Robocon; in that context, I discuss the way in which Buddhism and animism can inform audience reception.

CONTINUUM: ORGANIC BEINGS AND INANIMATE OBJECTS

Masahiro Mori discusses student experiences of robot competitions in relation to his educational philosophy on making and its relation to Buddhism (1999). Mori advocates that the students who participate in the building of competition robots can learn to understand the wholeness of the material environment, which includes humans. In one of his examples, a junior high-school student who worked with his classmates on their robot entry but could not attend the actual robot competition, made an intriguing comment: even though he was not at the event, the student felt his soul was in the robot, as were the souls of all the students that participated in the competition (Masahiro Mori 1999, 41). Mori notes it is not just this particular student who thinks this way but all the students who are involved in the school robot competitions. They all tend to think of robots as representing themselves in a spiritual way. Mori explains this phenomenon as typical of the situation in which a student is deeply engrossed in his or her robot-making: the student establishes a deep bond with the robot to the degree that he or she *forgets* him- or herself (1999, 70). This notion of forgetting the self in a creative activity, on one level, is not dissimilar to the concept I raised earlier of flow in regard to game-playing. One's engrossment in making a robot and participating in a competition could be explained through Turner's discussion, yet, on another level, Mori's raising of the forgetting of self significantly diverges from the logic of game-playing toward the domain of Buddhist thought. As Mori states, 'to forget oneself is to perceive oneself as all things' (1999, 71). His remarks reflect the Buddhist view that everything is connected with everything else in the universe. In line with this axiom, Mori explains in his earlier publication *The Buddha in the Robot* that 'selflessness means realizing that there is no fixed barrier between yourself and what is around you; it means knowing that you are linked with every other form of existence, animate and inanimate' (1981, 32). According to Mori, the human relationship with machines, exemplified by the situation of the student making a robot, is not at all one of master-slave, as Baudrillard remarks from a Western understanding, but one in which human and machine should instead be regarded as 'an interlocking entity' (1981, 179).

This view correlates to a taxonomic structure within the Japanese language. In Japanese, humans, animals, and plants are all conceived as things. While the Japanese word *seibutsu* carries the meaning of living creature, *dōbutsu* means animal, and *shokubutsu* means plant. In direct translation, *seibutsu* means living *thing*, *dōbutsu* means moving *thing*, and *shokubutsu* means planted *thing*. The character *butsu* shared by these Japanese words has a meaning close to that of an entity that lies outside the thing/object dichotomy, as it has been discussed in object studies. Humans, animals, plants, and inanimate nature are therefore all conceived *as part of* a world of material entities.[8]

In the next section of this chapter, to articulate Japanese subjectivity in terms of this close relation to the other, even where the other is non-human, I discuss what I call a reflexive anthropomorphism.

THE HUMAN, THE MONKEY, THE ROBOT, THE OTHER

Japanese anthropomorphism, as a dualism of self and non-self, should be understood as influenced by Japanese cosmological thought. According to anthropologist Emiko Ohnuki-Tierney, the Japanese mythological universe is based on 'symbolic oppositions' between humans and deities, inside and outside, where culture is the human way and nature is the divine world (1987, 129–30). These terms are placed in a mirroring relationship, where the latter term in the contrasting pair re-presents the former.[9] Japanese deities are understood as part of 'beings of nature . . .culturally defined to be "out there," beyond human society' (Ohnuki-Tierney 1987, 29, my ellipsis). In Japanese cosmology, the beings of nature include not only animals but also foreigners, those who are not of the Japanese people, who are of the 'inside', which is, by association, the human side. At the same time, the beings of nature on the 'outside' are mirroring the Japanese inside. For Ohnuki-Tierney, this mirrored structure is 'a projection of the dual qualities [of peace and violence] that the Japanese see in themselves' (1987, 134). This structure makes metaphoric figures of inversion extremely important.

In Japanese cosmology the monkey (the Japanese macaque) is one such figure, a symbol of inversion and ambivalence. In Japanese myth, the monkey is a revered mediator or messenger between deities and humans.

In the *kinsei* (Early Modern) period (1603–1868), the monkey was seen as transgressing boundaries between the human and the animal. To absorb this threat, the monkey was turned into 'a scapegoat, a laughable animal who in vain imitates humans' (Ohnuki-Tierney 1987, 6).[10] The same motifs and connotations are present in monkey performance. The monkey's foolish efforts to imitate humans are epitomised in the trained monkey's bipedal posture. This trick continues to be included in the repertoire of monkey performance in Japan to the present day.

Monkey performance reflects the paradoxical view of the monkey as an animal that is simultaneously the target of ridicule and the object of admiration. The collective sense of what it is to be a human can only be reaffirmed in conjunction with the monkey as a site of uncultivated nature/power. The monkey is able to present a form of the human, the cultured (that is, the Japanese), through a cultivated form of performance. There are two contradictory demands operating simultaneously here: the 'demand for maintenance of boundaries and demand for inter-categorical traffic' (Ohnuki-Tierney 1987, 156). In Japanese cosmology, the human world can only be conceived in relation to the non-human world, yet the former and the latter are in effect one and the same.

As a point of comparison, philosopher Emmanuel Levinas's phenomenological project, a product of his indebtedness to Edmund Husserl and Martin Heidegger, was intended to counter traditional metaphysics through which irreducible otherness is diminished and, in turn, a sense of the self secured and made concrete.[11] Levinas argued for one's ethical relationship and responsibility for 'other men [sic]', or 'the-one-for-the-other' (1991, 135–6). But in Levinas's theory there is still a hierarchical relationship between inside and outside: the former remains the dominant term. Indeed, ethics scholar David Gunkel critically points out that Levinas's anthropocentric view leaves unchallenged the position of the same, and 'the other is always and already operationalized as another human subject' (2012, 181). In traditional Japanese thought, on the other hand, the notion of the self as, in its essence, a composite entity of self and the world outside the self, has long been established. In a relation such as this, encompassing heterogeneous elements, the self and the world are inextricably linked. In Western popular thought and in Western philosophy, this way of thinking of the self, in which otherwise incompatible aspects coexist in a model of equality and cosmic balance, is not the norm—the centrality of the self is not questioned.

As I discussed earlier, the sense of the self in Buddhism is understood in relation to its surrounding environment, including objects. Like the view in Japanese mythology, in Buddhist thought the human is defined by the non-human. According to Mori, this non-human world may even include machines. Further, the non-human world may even exert a form of 'will' upon the human world that reflects its nature: machines such as cars don't do 'what you want them to do unless you do what they force you to do' (Masahiro Mori 1981, 177). This view recalls Steve Tillis's discussion of the reflexive relationship between the puppeteer and the puppet. Tillis argues that the puppet operator needs to 'show humility' in order to 'learn the movement potential of the puppet, and to allow for that potential to be realized' (1992, 162). While Tillis's comments nevertheless reflect that there is a level of control, Mori, on the other hand, applies a more Buddhist reflexivity and sense of equability to the human-machine relationship, arguing that 'there is no master-slave relationship between human beings and machines' (1981, 179).

ROBO-ONE involves contestants and officials, friends, supporters, and audience—all drawn together via these toy-like objects, the competing robots. Because its game structure unites everything within its scope, it is possible to speculate that ROBO-ONE may well facilitate the kind of traditional inter-category reflection between humans and inanimate objects with which its Japanese participants would already be familiar. In this kind of relation, aggression has no place.[12]

The participants of ROBO-ONE are aware that it is a playful event, but as I have mentioned with regard to Turner's notions of leisure and of communitas, it is 'serious play'—it is not designed to be funny, and it is certainly not intended to parody or mock the robots. Bacarobo, on the other hand, is an event that relies upon self-parody. How might Bacarobo create a sense of communitas among its participants and audience?

BACAROBO AND HUMOUR

Bacarobo is a comedy contest for robots that are designed to entertain the audience with laughter. This event was initiated by Nobumichi Tosa of Meiwa Denki (an art group that uses whimsical and bizarre electronic and mechanical instruments in its performances) in collaboration with Yoshimoto Creative Agency (a comedy/entertainment production company). It was held twice in Tokyo, in 2007 and 2008.[13] Literally meaning 'Idiot Robot', the Bacarobo competition uses strategies typical of the

Japanese television variety show context, and presents robots as integral players in humorous situations. To enter the competition, a participating robot must satisfy the following three principles of Bacarobo that appropriate the prescriptions of Isaac Asimov's well-known 'Three Laws of Robotics': (1) A Bacarobo has to be mechanical. It cannot just be a figurative model; it has to have an actual mechanical structure. (2) A Bacarobo has to be useless. Not intended as a functional robot deemed to be of use to society, its 'purpose' has to be trivial, frivolous, and worthless. (3) A Bacarobo has to make people laugh. Its ultimate utility is to bring laughter to viewers by startling them with interesting yet unusual movements, structures, and systems (Yoshimoto Creative Agency 2008). In the 2007 and 2008 events, the contestants, who were high-school students, renowned university professors, or professionals with engineering backgrounds, presented unusual robots that included a 'rubbish-bin' robot that repeatedly 'ate' rubbish and took it out of its 'mouth' again, and a 'dinosaur' robot that launched Frisbee discs from its mouth.

In Bacarobo, robots can become comedians when human performers and objects interact effectively within a coherent contextual matrix. More broadly, a mechanical object may function outside its usual understanding by enmeshing itself within the viewer's active imagination. What needs to be highlighted is that humour can facilitate the audience's imaginative capacity to accommodate something imperfect, strange, or unfamiliar, such as a robot, within a given social or performative context, that is, in a theatre. In that sense, I view Bacarobo in terms of clown theatre. The role of the clown can be discussed in terms of transformation and the socially ambivalent (Yamaguchi 2007). Clowning or buffoonery can be understood as a kind of performance or way of being that opens radically different ways of conceiving the world by turning the usual order of things upside down and altering one's sense of appropriateness through the structure of farce. What is created through this process is a transformative space that exists outside the everyday and that facilitates an inversion of the norms of societal value systems.

The most obvious aspect of Bacarobo's comedy comes from its robots' attempts to act like humans, or to mock human behaviour. This is amusing in the same way that the monkey is seen as funny in popular Japanese monkey performance when the monkey stands up on two legs or is dressed in tiny human clothes. The machines that ape or mock humans are the targets of laughter because they are clumsy, or imperfect mimics, just like the performing monkey. Bacarobo also trades in slapstick humour: robots do funny things, fall down, and get back up. The robot is the fool ('*baka*' in

Japanese), and it is the lower-status or inferior target of the event's broad, physical humour (Weitz 2016, 12). However, Bacarobo does not simply target the robot-clown for ridicule: the structure of Bacarobo also draws in, and makes targets of, its human participants in ways that I will discuss below.

Space, Community, and Audience

The winner of Bacarobo 2007 was a childlike robot that performed push-ups, which was seen as a cynical comment on the fitness boom at the time in Japan. Apparently, it kept doing push-ups even though its arm had fallen off and its neck was broken. The 2008 winner's prize was given to a long-necked robot with a young female face that was placed on a turntable and chased a viewer by turning itself. Occasionally, out of context, it would shake its head as if laughing. At the end of the 2008 contest, the judges explained that it won the award for exhibiting the greatest degree of practical uselessness. It is obvious that the winning machines were those that made fun of the usual view that robots should be practical and useful.

Despite its premise of the 'idiot robot', Bacarobo's humour was actually targeted not only at the machines but the human participants, and not just the robot's owner/creator. The contest itself took the form of a variety show. Its compère or MC (who was a professional comedian) and the judges (an artist, a writer, a film director, and a professor of media design) actively performed as a group of clowns who made fun of the human contestants presenting their 'stupid' robot inventions. They also laughed at the audience who had paid to see these useless robotic objects. At the same time, judges, audience, and contestants all considered the robots active performers. There was implicit agreement that they accepted the 'role' of these machines within the shared reality of the Bacarobo event. The robots were neither the non-human 'other' nor were they dangerous and menacing machines, the two ways that robots have tended to be presented in US robot competitions.

We can better understand how Bacarobo uses the physical situation of the event to create community by looking into the way Bacarobo organises the space between the performers and the viewers. Three groups of per-formers—MC, judges, and contestants with their robots—are demarcated; their performances occur in relation to those of each other. While the robot creators and their machines are viewed and commented upon by the other two groups, the MC and the judges are viewers of each other's performances. The audience watches this triangular exchange taking place on the stage. Importantly, in the manner of a vaudeville or variety show, the MC interacts with the audience with the house lights up. The MC makes comments that ridicule the audience, and its members are well aware that they are being watched. Thus it is clear that the audience is an integral part of the meaning created by the show's dynamics. This reflexive recognition is in a sense Brechtian: that is, the audience members become conscious of their own viewing situation. Their self-awareness in turn resets the 'front' of the theatre against the world outside the theatre. It connects the stage and the audience, forming a temporary consensus based on the fact that all present, including the audience, are 'fools' gathered around the strange and silly robots. The MC's use of the term 'we' to describe all the people at the event enhances the collectivity of what would otherwise just be an assembly of individuals.

I would like to briefly sum up my remarks concerning the way Bacarobo activates the theatrical space, incorporating what would otherwise be considered objects in performance. Sharing a consensual reality in a space designed to create such effects of 'communitas', the Bacarobo audience and participants realise that they stand together on the same ground. By viewing the human participants, including the audience, in relation to the robots, Bacarobo makes the robots members of the temporary Bacarobo community. The knowledge that all those present are members of a community of 'fools' generates a relaxed atmosphere of acceptance that extends to the strangeness and 'otherness' of the robotic machines. The message in the air may be something like: 'They (the robots) are not frightening, just stupid. But we are just as stupid (just as oblivious, just as inefficient) as the machines.' In proximity to the Bacarobo robots, the human participants' view of the robot as a practical and useful machine is temporarily put aside. The possibility for a different kind of relationship to the robot is implied.

We can understand the audience's affection for Bacarobo robots that fail, by noting, though it may seem a paradox, that while advanced robots epitomise technological and technical wonders for the Japanese, they also provoke a sense of uncertainty. Humanoids are regarded as engineering marvels with a long history of trial and error behind them. Yet, they are still very sensitive and fragile machines that risk failure. In terms of fragility, the humanoid demonstrations are similar to the *karakuri ningyō* shows. When the next-generation humanoids succeed in performing programmed tasks in their demonstration shows, the Japanese audience feels both a sense of surprise and of relief. The Bacarobo event can be seen to operate in the gap between desire engendered by the promised future of the new-generation robots and uncertainty. Bacarobo uses obvious humour, even ridicule, to take advantage of this matrix of response surrounding the idea of the socially desired humanoid robot.

Now that I have discussed how the set-up of Bacarobo's participant relationships is structured to create a sense of community, I will look more closely at the robot and its owner/creator.

CLOWNING AND FORMS OF COMEDY

In the transformative space of Bacarobo, it is not just the robots whose 'normal' roles are transformed or overturned. Bacarobo's strongest comic element is contributed by its human performers—the MC and the ludicrously costumed judges, who present themselves as clowns, as I indicated above. In the 2008 show, the clownish space of Bacarobo presented academics (two contestants—professor and researcher, respectively—and one judge, a professor) as 'strange' or even ridiculous, in contrast to the usual understanding of academics as intelligent and serious. This contrast can be expressed as something like this: while the professor and researcher came from respected universities supported by taxpayers' money, they produced such ridiculous machines. The professor/judge, whose academic background was also quite distinguished, was happy to be ridiculed and to show his fascination with the contestants' bizarre machines. Those who have respectable backgrounds, such as the university professor and researcher, are particularly targeted by the MC's mockery. This amusing contrast between respected adults and their 'stupid' inventions invites the audience to look twice at both the objects on the stage and the people associated with them: the machines are seen as doubly strange. In the world outside Bacarobo, it would not make sense to build deliberately non-functional robots, or for respected professionals to be wasting their time with them. However, within the *mise en scene* of the contest, the ridiculousness of the participating robots is paradoxically 'functional'.

The relation between humans and machines on the *Bacarobo* stage resonates with the relativism between actor and object in Jiří Veltruský's theory: the 'dialectic antinomy' between the human actor and the thing (1964, 90). Veltruský discusses, in theatre, 'a man [sic] can become a thing and a thing a living thing,' highlighting the transformative use of an object-like quality of the actor's presence and 'the action force' of the object (1964, 90 and 88). The styles and directions of particular performances determine the degree of the object-ness and behaviour of actor or object. In the case of Bacarobo, the relative 'stupidity' of the humans involved (the MC, the judges, and the contestants) drives 'the action force' of Bacarobo's machines. Thus in Bacarobo, the humans and the machines work side by side and are equally ridiculous.

This pairing structure of Bacarobo, its coupling of human sidekicks with robots, can also be discussed in terms of *manzai*, a form of Japanese stand-up comedy. *Manzai* comedy routines are performed in pairs: one comedian typically takes the role of *boke* (literally meaning a fool or an idiot) and the other is the *tsukkomi* (referring, in this context, to the contrasting role in the comedy duo). While the latter plays the 'normal one' or 'straight man', the former's role is to play the 'off-beat' character. The *boke* may be slow, vague, and stubborn. The *boke* may also be fast, energetic, and excessive. Regardless, the role of *boke* is that of the colourful character of the duo whose speech is full of misunderstandings and who makes mistakes. The *tsukkomi*'s role is to listen to or interrogate the *boke* if necessary and to point out the *boke*'s shortcomings. Part of the humour of this pairing comes from the fact that the *boke*'s nonsensical comments—typically childlike, blunt, honest, or outrageous remarks—are sometimes astutely observed, yet the *tsukkomi* has to maintain the upper hand as the voice of 'reason'. While the *tsukkomi* may express misgivings about his or her partner as part of their routine, there is, nevertheless, an unshakably strong bond between *boke* and *tsukkomi*.

The venues Rumine za Yoshimoto and Yoshimoto Mugendai Hall, where the two Bacarobo events were held, in 2007 and 2008, respectively, are prominent venues for *manzai* in Tokyo. As such, many contestants presented their machines in the form of *manzai*. Two robots might make silly 'conversation', for example. One of the 2008 contestants presented a robot as his *boke* partner. What was played out in these partnered presentations was the inextricable relation between the human and the technological. The robot performer and the human performer were each other's counterpart, as if they were the two sides of a coin. This relationship is also metaphorically a social one. Like typical Western ventriloquist dummy shows in which '[a] dummy saying the words in tones of voice that [a] ventriloquist never would, words that gain a certain independence from [the ventriloquist's] "real" life', the robot in Bacarobo might use vulgar, rude, and ill-mannered language in contrast to the calmness and politeness of the human performer (Goldblatt 2006, 55). This very contrast grants a performing 'body' and a social role to the robotic machine through voice and gesture that suggests interaction. The MC ridicules the human creator, even if he is the *tsukkomi*, as well as the robot *boke*, indicating the robot as inseparable from its human creator. In this sense, Bacarobo defines the performances of robots and their contestant-partners as collaborative.[14]

SATIRE AND *MODOKI*

It is necessary to note, however, that Bacarobo's humour is carefully controlled to fit mainstream tastes in comedy. There is no aggressiveness, 'camp', or deliberate perversion, as in the *otaku*'s fetishisation of manga/anime. Rather, Bacarobo is presented in the spirit of *modoki*, which, in the history of Japanese popular theatre entertainment, refers to a parody of an older, serious piece.[15] The imitative *modoki* practice is meant to be lightly humorous but not subversive. Mimesis, and thus mimicry, is the central subject of Bacarobo and its source of pleasure for the audience and participants; it is not concerned with social messages, as such, or political satire. The 'real' or 'serious' form that Bacarobo gently mocks, which is the practical humanoid robot, is kept well at bay. The appreciation of recognisable imitation in Bacarobo generates a common bonding effect, just as a circus audience would enjoy seeing elephants or tigers tamely performing, even though these animals could be menacing under other circumstances (Tait 2012, 2). Bacarobo's silly robots are already tamed, or castrated, and thus unlikely to frighten. Radical difference as such regarding Bacarobo strange machines is carefully kept hidden in Bacarobo's *modoki* approach.

As I have discussed in the introduction to this book, the ancient meaning of *matsuri* is as an entertainment for *marebito*, referring to foreign gods or messengers (Orikuchi 1991, 22). '*Geinoh*' practitioners, artists of popular and folk entertainments, were the dancers and singers for the ancient *matsuri* (Orikuchi 1991, 22). Bacarobo is a contemporary, secular *matsuri* event, with 'weekend' *geinoh* participants. In the same way that traditional *matsuri* incorporates all of its participants in a temporary community, Bacarobo absorbs all of its participants and spectators through the medium of the robots, rendered as fools, into the easy humour of *modoki*. Bacarobo embodies 'the phenomenon of laughter' that transforms audience members into crowds, in which 'people's responses aren't always governed by their conscious mind' (White 2013, 132).

It is important to be reminded that, as drama studies scholar Eric Weitz discusses, laughter is a 'technology' in a Foucauldian sense, a mode through which 'we surrender to a state of bodied submission' (2016, 69). While laughing at others keeps them in line, it is 'the threat of being laughed at that keeps *us* in line' (Weitz 2016, 71, original emphasis). Social appropriateness is inscribed in laughter, as philosopher Henri Bergson points out: 'Laughter is, above all, a corrective. Being intended to humiliate, it must make a painful impression on the person against whom it is directed' (1935, 197). Laughing with others generates a sense of bonding with them, which, in turn, excludes 'those who are not in the group', and this laugh bonds people in an 'involuntary' way (White 2013, 131). Underneath the surface of comedy, Bacarobo executes a similar operation of assimilation.

This chapter has looked into the transformative and mediating effects of particular *mises en scène* at the sites of ROBO-ONE and Bacarobo. Such robot contests encourage and facilitate a positive view of robots. They are, in effect, successful public relations exercises for the larger Japanese robotics industry, improving the public perception of the robot by staging a kind of 'three-legged race' between humans and robots in a family-friendly atmosphere that is not unlike that of the *matsuri*.

In the final chapter, I look into the use of interactive social robots in aged-care institutions. If Bacarobo relied upon its participants' awareness of the joke, the residents in these institutions may not be participating in quite the same way. Ethical issues arise concerning elderly patients' cognisance of their interaction with robots in highly organised activities that are intended to elicit affective response.

NOTES

1. ROBO-ONE, as an organisation, indicates its name in English in capital letters.
2. The oldest ongoing robot competition in Japan is All Japan Micromouse Contest, in which an autonomous, mouse-like robot on wheels moves to get to set positions in a maze. Inspired by the original American event in 1977, this contest began in 1980.
3. Other robot competitions for students include All Japan Robot-Sumo Tournament, held annually since 1989, in which two non-anthropomorphic robots try to push each other out of a ring, as in sumo; and Kawasaki Robot Festival, which began in 1994 and is a battle competition of radio-controlled, creature-like, robotic vehicles with arms that can be used to throw an opponent to the outside of a square ring.

4. It is relevant to note that according to computer scientist Hitoshi Matsubara, a key member of the RoboCup committee, there is the view that RoboCup lacks the entertainment aspects of a spectator sport compared with 'real' soccer games by humans, due to the absence of individuality and character in its robot 'athletes' and that it also lacks the human aspects of real soccer matches, such as players making deliberate fouls (2001, 32–42).

5. ROBO-ONE has been evolving since its inauguration. Currently, apart from the main competition, ROBO-ONE, there are two partner competition forums: ROBO-ONE Light and ROBO-Ken. ROBO-ONE Light (which started in 2010) is designed for beginners, using authorised models of commercially available robot kits, such as Kondo Kagaku's KHR series. ROBO-Ken (which began in 2013) is a tournament of one-armed robots performing kendo, Japanese swordsmanship.

6. For Sena, ROBO-ONE is a realisation of the manga/animation *Puraresu Sanshirō* in the early 1980s, which is a story about a boy protagonist and his model wrestling robot and their fights in robot battle tournaments (2004, 516).

7. The TV series of 'caged' fight contests of unmanned robot vehicles were aired in the UK (*Robot Wars* from 1998 to 2004) and in the USA (*BattleBots* from 2000 to 2002, a new season from 2015).

8. This structure can also be seen in relation to the study by Émile Durkheim and Marcel Mauss on non-Western classifications of objects, in which they found that these systems 'make intelligible the relations which exist between things' (Durkheim and Mauss 1963, 81).

9. This concept can be contrasted to the Lacanian 'mirror stage' through which a child develops a sense of distinction between the self and the other. The 'mirroring relation' in Japanese cosmology encourages the blurring of the two.

10. For a similar idea of the monkey as imitator of the human in the European context, see Weston (2009).

11. Referring to Levinas, communication study scholar Eleanor Sandry argues for an asymmetrical, yet dynamic and fluid relationship between humans and non-humanoid robots 'by virtue of their different forms' (2015, 115).

12. In a sense, the contrast between ROBO-ONE and its American counterparts parallels the difference in the directions of robotics

research in Japan and the USA. While a large proportion of the USA's government funding is directed to robotics research for military purposes such as drones, Japanese robotics research focuses on non-military, that is, corporate and civilian, use (Asada 2011, 6). It is understood that this 'pacifist' direction was determined in 1950 when the Science Council of Japan (SCJ), a statutory body that liaises between the government and the scientific community, disallowed scientific studies for military purposes (*The Yomiuri Shimbun* 2016). Shigeo Hirose confesses that he was wary of criticism when he published his research on robots for mine clearance as the first Japanese roboticist to do so (2011, 4). However, in 2013, the Abe government adopted a new defense policy to direct funds into 'dual-use' technologies that have both military and civilian applications. The science community is currently very divided on this issue (*The Yomiuri Shimbun* 2016). While it is foreseeable that Japanese robotic technologies may be developed for military use in the future, the particular Japanese relationship to robots, as this book discusses, would continue in one way or another.

13. Bacarobo (as The Bacarobo Stupid Robot Championship) was also held in Budapest in 2010 and 2011, involving international participants. I exclude these events from my examination, as this book focuses on robot events in Japan.

14. In more recent events of Meiwa Denki in 2014 and 2015, Tosa uses Pepper (Aldebaran Robotics), one of the most advanced commercially available communication robots. This humanoid (1.2 metres in height, on wheels) is equipped with human-like, expressive arms. While Pepper's speech function is sophisticated, the communication between the robot and human performers is still not smooth and its timing when responding to human performers is often out of sync. The human performers have to work to make up for the gaps. The same applies to the *manzai* by PaPe-jiro, a comedy duo between comedian Zenjiro and NEC's communication robot PaPeRo, a 40 cm high talking head.

15. The original practice of *modoki* refers to the imitative response by Japanese indigenous gods, repeating and copying the words and behaviour of the foreign god, *marebito* (Orikuchi 1991, 164).

REFERENCES

Azusa, M. (2009, November 2). Dai jū rokkai ROBO-ONE in Toyama repōt [A report on The 16th ROBO-ONE in Toyama]. *Robot Watch*. http://robot.watch.impress.co.jp/docs/news/326137.html. Accessed on 01-15-2010.

Asada, M. (2011). Robotto to saiensu ga michibiku ugoki, katachi to shikō no aratana kagaku (A new area of scientific research concerned with movement, form, and thought, led by robot science). *Bessatu Nikkei Saiensu, 179*, 4–6.

Baudrillard, J. (2005). *The system of objects* (J. Benedict, Trans.) (2nd ed.). London/New York: Verso.

Bergson, H. (1935). *Laughter: An essay on the meaning of the comic* (C. Brereton & F. Rothwell, Trans.). London: Macmillan and Co.

Borenstein, J., & Pearson, Y. (2012). Robotic caregivers: Ethical issues across the human lifespan. In P. Lin, K. Abney, & G. A. Bekey (Eds.), *Robot ethics: The ethical and social implications of robotics* (pp. 251–265). Cambridge, MA: MIT Press.

Caro, M. (2008). *Astroboy in Roboland*. HD Video. Paris: Les Films d'Ici.

Durkheim, É., & Mauss, M. (1963). *Primitive classification* (R. Needham, Trans.). London: Cohen & West.

Goldblatt, D. (2006). *Art and ventriloquism*. London/New York: Routledge.

Gunkel, D. J. (2012). *The machine question critical perspectives on AI, robots, and ethics*. Cambridge, MA: MIT Press.

Hotta, J. (2008). *Hito to robotto no himitu* [The secret concerning robots and humans]. Tokyo: Kōbunsha.

Hirose, S. (2011). *Robotto sōzōgaku nyūmon* [Introduction to a study on robot creation]. Tokyo: Iwanami shoten.

Latour, B. (1996). On interobjectivity (G. Bowker, Trans.). *Mind, Culture, and Activity, 3*(4), 228–245. doi:10.1207/s15327884mca0304_2.

Levinas, E. (1991). *Otherwise than being, or, beyond essence* (A. Lingis, Trans.). Dordrecht/Boston/London: Kluwer Academic Publishers.

Matsubara, H. (2001). Robo kappu no yume [A dream of RoboCup]. In H. Matsubara, I. Takeuchi & H. Numata (Eds.), *Robotto no jōhōgaku: 2050 nen wārudo kappu ni katsu* [Robot information study: Winning the World Cup in 2050] (pp. 7-66). Tokyo: NTT Shuppansha.

Maxwell, I. (2008). The ritualization of performance (studies). In G. St. John (Ed.), *Victor Turner and contemporary cultural performance* (pp. 59–75). New York/Oxford: Berghahn Books.

Mori, M. [Masahiro]. (1981). *The Buddha in the robot* (C. S. Terry, Trans.). Tokyo: Kōsei Publishing.

Mori, M. [Masahiro]. (1999). *Robo kon hakase no monotsukuri yūron* [Dr Robocon on the art of making]. Tokyo: Ōmusha.

Mori, M. [Masahiro]. (2014). *Robotto kōgaku to ningen: Mirai no tameno robotto kōgaku* [Robotics and the human: Robotics for the future]. Tokyo: Ōmusha.

Ohnuki-Tierney, E. (1987). *The monkey as mirror: Symbolic transformations in Japanese history and ritual.* Princeton: Princeton University Press.

Orikuchi, S. (1991). *Nihon geinōshi rokkō* [The history of Japanese entertainment: Six lectures]. Tokyo: Kōdansha.

Pyne, L. (2014, September 30). The day we brought our robot home. *The Atlantic.* http://www.theatlantic.com/technology/archive/2014/09/the-day-we-brought-our-robot-home/380891/. Accessed on 10-03-2015.

Sandry, E. (2015). *Robots and communication.* New York: Palgrave Macmillan.

Sena, H. (Ed.). (2004). *Robotto Opera (Robot Opera: An anthology of robot fiction and robot culture, original English title).* Tokyo: Kōbunsha.

Tait, P. (2012). *Wild and dangerous performances: Animals, emotions, circus.* Houndmills/Basingstoke/Hampshire/New York: Palgrave Macmillan.

The Yomiuri Shimbun. (2016, July 1). Gunji, Minsei Kakine Koeru Gijutsu: Bōeiyosan, Hantai No Daigaku Mo (Technologies that cross the wall between military and civilian use: A defense budget and opposing universities). Yomiuri Online. http://www.yomiuri.co.jp/osaka/feature/CO022791/20160701-OYTAT50015.html. Accessed on 07-30-2016.

Tillis, S. (1992). *Toward an aesthetics of the puppet: Puppetry as a theatrical art.* New York: Greenwood Press.

Turner, V. (1982). *From ritual to theatre: The human seriousness of play.* New York: PAJ Publications.

Turner, V. (1988). *The anthropology of performance.* New York: PAJ Publications.

Veltruský, J. (1964). Man and object in the theater. In P. L. Garvin (Ed.), *A Prague school reader on esthetics, literary structure, and style* (pp. 83–91). Washington, DC: Georgetown University Press.

Weitz, E. (2016). *Theatre and laughter.* London/New York: Palgrave Macmillan.

Weston, H. (2009). Fables and follies: Florian's 'The monkey showing the magic lantern' and the failure of imitation. In A. Satz & J. Wood (Eds.), *Articulate objects: Voice, sculpture and performance* (pp. 47–62). Oxford/New York: Peter Lang.

White, G. (2013). *Audience participation in theatre: Aesthetics of the invitation.* Houndmills/Basingstoke/Hampshire/New York: Palgrave Macmillan.

Yamaguchi, M. (2007). *Dōke no minzokugaku.* Tokyo: Iwanami Shoten.

Yoshimi, S. (2008). *Toshi no doramatrugī: Tokyo sakariba no rekishi* [Dramaturgy of the city: History of entertainment districts in Tokyo]. Tokyo: Kawade shobō.

CHAPTER 8

Robots that 'Care'

As I discussed in Chap. 3, the government-led Humanoid Robotics Project (1998–2003) encouraged roboticists to develop robots that could contribute to the management of an ageing population. The government agency New Energy and Industrial Technology Development Organisation (NEDO) has identified areas in which robots can assist at hospitals and care facilities, referring to, for example, 'care' robots, 'pet' robots, 'nurse assistant' robots, and 'dementia prevention' robots, to name a few (2009, 126–131). Government promotions indicate that these next-generation robots would assist the elderly and their carers in performing everyday tasks, in monitoring the patients' health, and in communication activities. In this chapter, I look into the use at aged-care institutions in Japan of interactive robots of zoomorphic appearance, which are presumed to communicate with humans with sound and/or movement. This chapter extends my discussion of social 'performance' in Japan in regard to next-generation robots, even in settings outside the theatre or staged situations.

In particular, I examine the discussions on experimental sessions with communication robots at aged-care facilities in Japan, focusing on the question of performance. The engagement of dementia patients with therapeutic robots is seen positively by the proponents of these machines. Critics, on the other hand, look at the use of the social robot in aged care in terms of duplicity and control. The communication that is encouraged between these patients and the therapy robots is seen as fake and contrived. Social psychologist Sherry Turkle, known for her studies in human-technology interaction, discusses what she sees as the deceptive nature of

© The Author(s) 2017
Y. Sone, *Japanese Robot Culture*,
DOI 10.1057/978-1-137-52527-7_8

social robots with reference to performance in its colloquial sense, meaning an act that is artificial, not real (2011, 6). Taking my cue from Turkle's concerns, this chapter discusses the mechanisms of exchange in these activities with robots. I elaborate upon Turkle's discussion of performance in terms of Richard Schechner's concept of 'make-belief', an action that is staged as 'real', blurring the distinction between 'what's real and what's pretended' (2013, 43). An inclusive and friendly atmosphere is created in care facilities in order to accentuate robots' affective appeal. The robots' design and the nature of their movement are, of course, designed to trigger interest and empathy. Various measures—such as the use of keywords intended to spark a patient's memory or imagination—are deployed by carers working with these robots in order to encourage patient participation. In essence, a transformative atmosphere is created with the performance of zoomorphic robots at its centre. Robot sessions for patients rely upon the activation of robots' inviting design features in relation to these dramaturgical considerations for their effective presentation to patients.

The final section of this chapter discusses the case of private robot owners, mostly middle-aged and older, who are attached to their 'pet' AIBO (robot dog). These owners can be regarded as forming a particular fan subculture for the AIBO robot, but one that parallels, in its evocation of affective response, the use of AIBO in aged-care institutions. This correlation reflects a unique, and disturbing, characteristic of Japanese robot culture.

THE SOCIALLY ASSISTIVE ROBOT

The robots that I discuss in this chapter are the interactive zoomorphic robots: Paro, which looks like a baby harp seal, and AIBO, Sony's iconic robot dog.[1] Paro and AIBO have been used as 'socially assistive' robots in numerous studies on aged care in the US, Australia, and Japan. (Kanamori et al. 2003; Tamura et al. 2004; Wada and Shibata 2007; Banks et al. 2008; Rabbitt et al. 2015; Moyle et al. 2015).[2] The 'socially assistive' robot is a category of next-generation robot that intersects with the types of robots known as 'social robot', 'service robot', and 'assistive robot'. In Japan, the term 'social' robot generally describes robots designed to interact with people, such as those I discussed in an earlier chapter by Hiroshi Ishiguro.[3] Social robots in Japan can come in various forms and sizes, be autonomous or semi-autonomous, movable or static humanoid, android, or zoomorphic, as I have indicated earlier. 'Service' robots in the professional domain can include delivery or cleaner robots for large public spaces or office buildings that look like mobile machines.[4]

Assistive robot refers to a robotic machine that assists 'people with physical disabilities through physical interaction' for domestic and institutional use (Feil-Seifer and Mataric 2005, 465). This type of robot includes robotic devices that help people with physical disabilities with eating or with transportation between bed and wheelchair.[5] The term 'socially assistive robot' delineates a subcategory within assistive robots, which refers to robots that provide assistance to humans, through interaction with verbal and/or physical gesture but without physical contact (Feil-Seifer and Mataric 2005, 465). Socially assistive robots have been used as companions or guides for the elderly and children in pilot implementation studies and, more recently, in randomised, controlled studies (RCT, quantitative method for scientific studies).[6] According to social scientists, while issues in terms of ethics, cost, safety, and the impact upon job losses remain to be addressed, the positive potential of socially assistive robots is acknowledged. (Sparrow and Sparrow 2006, 156; Borenstein and Pearson 2012, 254; Sharkey and Sharkey 2012, 277; Rabbitt et al. 2015). The positive effects for elderly patients, they claim, include an increase in positive mood and a healthier immune system, while alleviating stress and feelings of loneliness. However, methodologies to measure such effects have been questioned as well (Broekens et al. 2009).

Japanese researchers see the potential in the use of socially assistive robots for aged care, based on the data obtained from their studies, although there are unresolved issues such as high cost and technical complexity of implementation, as well as the lack of agreed methods for measuring their effectiveness.[7] It is important to point out, as cultural anthropologist Shawn Bender notices, that few critical or challenging philosophical and ethical arguments on socially assistive robots have been put forward by Japanese scholars in the humanities and social sciences (2012, 648). Studies on the socially assistive robot in Japan have so far been dominated by introductory and promotional discussions by roboticists who have had direct involvement in the development of such machines. In other words, studies on socially assistive robots for the care of the elderly in Japan are most often led by university-based and corporate researchers on robotics who describe their own projects and argue for the potentials of their robot prototypes and products for such use. Naturally, many of their studies are published in *The Journal of the Robotics Society of Japan*, which is run by the Robotics Society of Japan, the peak academic body for university robotics researchers, and *Robotto* (robot), an in-house bulletin by The Japan Robot Association that represents the robot manufacturing industry. The widespread justification, even in these publications, for the use of socially assistive robots in aged care is often based on a commonly held view that the Japanese prefer robots over human helpers, especially foreign health workers (Robertson 2007, 372).[8]

More specifically, the socially assistive robot for aged care in Japan has been discussed as an alternative not only to human workers but to 'animal therapy' (Hamada et al. 2003; Suzuki et al. 2009; T. Shibata 2010; Yokoyama 2012).[9] While scientists are not always in agreement, animal use in therapeutic contexts at home or in institutions is said to have a positive impact on patients' social, emotional, and cognitive abilities. Animal therapy can be categorised into two groups: 'animal-assisted therapy' that is associated with particular objectives often set by medical practitioners and 'animal-assisted activity' through which patients spend unstructured time playing with animals. Socially assistive robots are used in Japan mainly as 'robot-assisted activity'. Their perceived benefits over animals are their cleanliness, as animals can carry germs; safety, since animals can bite and scratch patients; stress-free use and maintenance, avoiding the need to address animal welfare needs; programmable, because animals would need to be trained; and they are owner-free, as an animal owner would need to be present at a care facility (Yokoyama 2012, 600). Also, the sad experience of an animal's death can be avoided when using a robot (Hamada and Sano 2011, 595).

The development of interactive robots that would work in close proximity to humans in domestic environments has been supported by the Japanese government for more than a decade. The ethical issues of using robots in aged care may be minimised in relation to the pressing need to address the economic and social problems presented by a shrinking and ageing population. After years of predictions, the 2015 national census indicates that the Japanese population decreased by nearly one million between 2010 and 2015. It is the first substantiated record of population decline since the 1920s (The Statistics Bureau, Ministry of Internal Affairs and Communications 2016a). A 2013 statistic indicates that one in four Japanese people is over 65 years old (25.9 %) and one in eight is over 75 years old (12.5 %) (The Statistics Bureau, Ministry of Internal Affairs and Communications 2013).[10] Instead of an approach that focuses on the development of humanoids that can replace caregivers, per se, there has been a recent emphasis on more pragmatic approaches such as the use of less expensive assistive robots at sites where nursing is required (The Ministry of Economy, Trade and Industry 2015b, 68). The government encourages the development of robotic devices ready for commercialisation in areas '"for transfer support (indoors)" and "for watching over dementia patients (for institutions)"' (The Ministry of Economy, Trade and Industry 2015b, 69, original English).[11] In the case of socially assistive robots, small desktop-stationed social humanoids, which are linked to the Internet and to telecommunication and home electronic devices, are, for example, being promoted at trade fairs, such as the iREX (International Robot Exhibition) 2015 as multifunctional information terminals.[12]

In contrast, Western researchers from disciplines outside robotics have critically analysed the use of socially assistive robot for aged care with direct and indirect references to the Japanese cases (Sparrow and Sparrow 2006; Robertson 2007; Turkle 2011; Borenstein and Pearson 2012; Sharkey and Sharkey 2012; Sorell and Draper 2014; Rodogno 2015). The general concern with the use of assistive robots in aged care is that these machines can be seen as a 'technological fix' that is a 'compensation' for a deficit in carer numbers (Borenstein and Pearson 2012, 260 and 262). More specific concerns include a decrease in human contact, the mistreatment of elderly patients as objects, and the loss of control over one's own life (Sharkey and Sharkey 2012, 30). There is a strong view that the idea of the robot companion and its presumed benefits are 'premised on people believing that robots are something that they are not' (Sparrow and Sparrow 2006, 148). The general view in the West concerning the use of socially assistive robots in aged care is that it is premised upon deception.

Turkle's theorising of her study of subjects in interaction with social robots reflects upon the complexities of the process of communication between humans and robots. Turkle was surprised to discover that her subjects in the USA—young children, adults, and the elderly—felt they could communicate with these machines, even with a degree of affection. Turkle points out her subjects' readiness to regard robots 'not only as pets but as potential friends, confidants, and even romantic partners' (2011, 9). What these machines actually do, according to Turkle, is tap into human vulnerability. When people are lonely and desperate for interaction, these substitutes can be good enough. However, such interactions are constituted merely by superficial appearance, which is, for Turkle, precisely what is offered by the social robots (2011, 72). When humans bond with a robot, they see it as responsive: 'It becomes "alive enough" for relationship' (Turkle 2011, 18). For Turkle, this is anthropomorphism through '*a show* of friendship', and, in 'the robotic moment, the performance of connection seems connection enough' (2011, 8–9, my emphasis). This relation replaces human contact, and it is exploitative and convenient. For Turkle, Japan provides a prime example of an enthusiasm for robots leading society in the wrong direction.

What needs to be noted is that the driving force for the use of socially assistive robots for the elderly comes from Japanese roboticists and the Japanese robot industry, with support from the government, and their clear objective is to make this area a growing and commercially viable field for next-generation robotics. Varied interests have converged upon the use of the socially assistive robot in aged-care facilities. As I have suggested, these interest groups include not only the robot manufacturing industry and academics in robotics, but also scholars and experts in other fields, including psychology, nursing, and the social sciences (Rabbitt et al. 2015, 42; Yokoyama 2012, 600; Sparrow and Sparrow 2006, 156). A critical reading of this situation is that underneath shared interests, there is a common view that old people are seen 'as problems, or as objects of study' and the use of the socially assistive robot presents 'technical solutions' to the complications they create (Sparrow and Sparrow 2006, 156). There is the obvious risk that experiments with the cognitively impaired elderly in Japan could be carried out with 'no clear benefit to the affected parties other than a stimulation of the robot-producing industry' (Kaerlein 2015, 86).

Turkle's comments above may suggest that social robots have become the norm in Japan. While the integration of social robots in Japanese daily life is conceived as a national project, it is still in its development phase and not yet at all a common phenomenon to see them in homes or institutions (Bender 2012, 647). Nevertheless, Turkle's notion of performance at 'the robotic moment' is useful. It is my contention that while Japanese roboticists and related experts discuss the value of human-robot interaction, their desideratum, which is never overtly stated, seems to be a state of engagement in which the human participant forgets or rather, cognitively knows but affectively forgets that he or she is talking with a robot. This is Turkle's idea of performance at 'the robotic moment', and it is imbedded in the design philosophy and implementation of Japanese socially assistive robots.

To explain these points further, I first examine the designs of AIBO and Paro in relation to the contexts in which they are deployed. These mimetic robots can colourfully mediate humans' interaction with them when they are operated in an effective *mise en scène*.

AIBO'S DESIGN

To start to understand how robotic machines like AIBO and Paro are utilised, I refer to Bender's ethnographical observation on 'psychological therapy using robots' at an aged-care facility in Japan (2012, 645). Bender attended one-hour activity sessions that featured communication robots such as Paro, NeCoRo, AIBO, and ifbot. The sessions, in which patients sat around a table, were facilitated by the staff members and a team of roboticists. Bender's study revealed that contrary to the common under-standing that robots simply communicate with humans through verbal and gestural means alone, the robots in the sessions were always present with human facilitators. The robots were regarded not as 'an end in themselves' but as 'a means through which nursing home residents could interact with each other and with other staff members' (Bender 2012, 647). The facili-tators' involvement was in actuality central to this interaction. The level of the facilitators' ability to use the robots determined the effectiveness of the robot therapy sessions. These outcomes are consistent with those of Japa-nese researchers in rehabilitation studies: they discuss robot therapy as a non-everyday stimulus that requires mediators to motivate the subject (Kato 2012, 605), emphasising the importance of an experienced mediator (Yokoyama 2012, 601; Kimura 2012, 636; Kameda 2015, 22). Sena points out that a challenge for the designer of anthropomorphic and zoomorphic robots is to maintain the user's interest in the machine for an extended period of time (2008, 385). An experienced facilitator is understood as necessary in this regard.

The use of robots for therapy involves a kind of theatrical structure: with the facilitator as MC, robots as performers, and patients as the audience. The robots become a 'shared topic of conversation', functioning as a kind of prop for 'talk therapy' (Bender 2012, 647). For psychiatrist and researcher Akimitsu Yokoyama, the role of the human facilitator in robot-assisted activity is very important, as socially assistive robots make the patients aware of the people around them, unlike animals, which become the focus of the elderly person's attention (2012, 601). Yokoyama goes on to describe robot-assisted activity as '*gokko asobi*' (children's play) because the elderly person knows that robots are not alive, yet they are able to treat them as if they are living creatures (2012, 601). According to Yokoyama, a capable facilitator would be able to generate a relaxed and inviting atmosphere for 'play' that would facilitate improvement in the patients' ability concerning his or her sociability, empathy, and awareness (2012, 601).

For the role of MC/facilitator, certain verbal techniques to 'direct' patients are recommended. For example, in relation to Paro, its creator Takanori Shibata discusses how references to a patient's previous pet or favourite animals can generate opportunities for conversation with the patient (Shibata and Yukawa 2014, 37). Shibata describes a situation where a staff member asked a female Alzheimer's patient to 'look after' Paro by referring to the robot with the name of the patient's dog (Shibata and Yukawa 2014, 37). Kazuyoshi Wada, a roboticist who collaborates with Shibata, recommends that care facilitators provide specific verbal cues to patients in order to maintain such contextual meanings in regard to Paro (2012). For example, it would be better to say, 'Paro will be waiting for you', when a patient has to go to the toilet. When the patient returns, facilitators' comments such as 'Paro looks happy' or 'Please play with Paro again' would serve to encourage the patient to reengage with the robot. The facilitator's comments are intended to transform a piece of robotic machinery into a 'creature' of some sort, in a form of play with a toy. While children will participate with enthusiasm in games requiring a 'suspension of disbelief', the cognitively impaired elderly need to be encouraged. In dealing with a group of such elderly patients, the role of the facilitator may become coercive.

The design strategy for socially assistive robots is to seek response. In order to clarify their mechanisms of response, I refer to Donald Norman's three levels of industrial design: visceral, behavioural, and reflective. Design at a visceral level is about 'the initial impact of a product, about its appearance, touch, and feel' (2004, 37). At the behavioural level, use and experience of a product's 'function, performance, and usability' is considered (Norman 2004, 37). Unlike the phenomenological operations of the first two levels, 'thought and emotions' are engaged through '[i]nterpretation, understanding, and reasoning' at the reflective level (Norman 2004, 38). These three concepts can be used to examine the seductive effects of mimetic communicative robots such as AIBO, which I will treat first, and Paro.

Astrid Weiss, Daniela Wurhofer, and Manfred Tscheligi, a team of researchers on human-computer interaction, discuss their experiment to see differences in reaction and response between adults and children when interacting with AIBO, using Norman's schema (2009). The children were quick to respond to AIBO when introduced to it (at the visceral level of engagement), uttering comments such as 'Oh, what a nice dog', 'That's so cool', and 'May I play with it' (Weiss et al. 2009, 246). The children's behavioural level of engagement was manifest in the form of play with AIBO, resulting from the children's initial impressions of it at the visceral level of engagement. Their reflective level of engagement was evidenced by indications of attachment to the robot, such as seeing it as a fellow playmate (Weiss et al. 2009, 246).

The children's responses are not surprising as, given this study by Weiss, Wurhofer and Tscheligi, AIBO's design seems firmly consolidated at the visceral and behavioural levels. While there were a few changes in appearance in the series, AIBO maintained a visually appealing quality, with a cute yet sleek and futuristic look that suggested a sort of 'intelligence' (AIBO's project leader, Tadashi Otsuki quoted in A. Kubo 2015, 212).[13] With the sound of its electric motors, which highlight its machine quality, it walks, sits with its back legs bent and the front legs straight like a real dog, stretches, and wags its tail. The later series of AIBO were equipped with a touch sensor on its back, and it understood (translated) different kinds of touch, as in strong touch, such as smacking, or a soft touch, such as caressing or a pat on the back. Responding to these stimuli, AIBO 'expresses' happiness, sadness, anger, fear, or disgust through a combination of the light display on its face cover and the gesture and arm motion of 'banzai' (two arms raised), dropping its head, or moving backward. Norman states that some owners of the AIBO 'believe that their personal AIBO recognizes them and obeys commands even though it is not capable of these deeds' (2004, 194). Moreover, there are a few patterns for each feeling. For example, when expressing anger, AIBO's face cover might display a red colour or it might flap its front legs, followed by 'sulky' gestures (Yamaguchi 2006, 242). It is possible to imagine that the children in the experiment by Weiss, Wurhofer, and Tscheligi might have recognised AIBO's movements that suggest naughtiness or coyness. Visual and behavioural 'tricks' such as these would help to establish an illusion of exchange with AIBO for children.

AIBO also embodies the design concept that encourages in its user a slightly more complex reflective engagement in conjunction with the visceral and behavioural levels. Toshitada Doi's comments on the philosophy of the AIBO project in relation to Asimov's 'Three Laws of Robotics' are indicative in this regard. While it follows the first law of refraining from harming humans, Doi states that AIBO should be able to disobey orders from time to time, contrary to the second law. As for the third of Asimov's laws, which is about self-protection, AIBO should release 'stress' by 'whingeing' from time to time (quoted in A. Kubo 2015, 209). Indeed, AIBO is designed to resonate with the concept of the 'useless' robot as opposed to the usefulness of industrial robots (Yamaguchi 2006, 229). In order to strengthen the illusion of the robot's aliveness, certain 'scenarios' are pre-programmed. For example, when AIBO falls, it gets up clumsily, rolling its body left and right slightly, which gives its owner the sense that his or her adorable AIBO is trying to get up (Yamaguchi 2006, 243). The AIBO design choreographs the machine's limited expressions, eliciting the owners' imaginary scenarios.[14] For example, a few walking patterns—from clumsy walk to steady gait—are programmed into AIBO, which generate a sense of 'growth' (Sena 2001, 138). Even though 'AIBO, the Sony robot dog, has a far less sophisticated emotional repertoire and intelligence than Kismet', according to Norman, it has 'proven to be incredibly engaging to its owners' (2004, 194). Owners who have spent a long time with their robot dogs may recognise unusual patterns in AIBO's movements but see these idiosyncrasies as reflective of their own AIBO's uniqueness, even though, in fact, they indicate technical problems (A. Kubo 2015, 225). Certain 'quirks' are seen by the owner as the robot's 'possessing traits characteristic of persons' (Borenstein and Pearson 2012, 253). (I will discuss a case that illustrates AIBO owners' attachment to their individualised robots shortly.)

Due to its technological limitations, it is difficult for an interactive social robot such as AIBO to sustain an ongoing and meaningful exchange with an adult human user without the robot being seen as gimmicky. Sony embedded appealing phrases in the user information for AIBO that would compensate for technical shortfalls, playing on the owner's expectations of the futuristic gadget and their willing emotional attachment to it. The operation manual for AIBO states that: 'Your AIBO will never die and it can't be reset like a video game. AIBO will remain your partner always' (quoted in A. Kubo 2015, 207, original English). Sony names the customer help centre the 'AIBO Clinic', in order to differentiate it from other consumer electronic products (Yamaguchi 2006, 234). As much as the design of AIBO robot dogs, the context around it works to elicit a user's emotional attachment, a strategy that has been critiqued with terms such as 'synthetic emotion' (Turkle 2011, 123) or 'artificial emotion' (Bekey 2012, 30). Humans have a natural tendency for 'anthropomorphic sentimentalization', and, hence, 'production and consumption of such technology should be morally permissible', according to social scientist Raffaele

Fig. 8.1 Robot seal 'Paro', OSAKA, Japan—Daiwa House announces it will lease robot seal 'Paro' for free in the areas affected by the Great East Japan Earthquake, 2011 © Kyodo News (Courtesy of Kyodo News International)

Rodogno (2015, 9). AIBO takes full advantage of people's anthropomorphic tendencies. This anthropomorphism arises in the same way as that directed toward other robot objects in the Japanese context, as I have discussed throughout this book; it is also the case, as I have indicated, that such attachments can be redirected toward nationalistic and governmental aims. The situation is no different concerning the care of the cognitively impaired elderly in aged-care institutions, and I will return to such questions later. For the moment, while AIBO and Paro are both used in institutions, I will focus on Paro because its use reveals ethical grey areas in human-robot interaction more starkly (See Fig. 8.1).

COMMUNICATION WITH PARO

Paro embodies a much simpler design in terms of appearance and mechanical functionality than does AIBO. However, complex strategies inform Paro's design, because it is produced for elderly care situations, in long-term use.[15] Shibata explains the reasons behind designing Paro as a baby seal, as opposed to a puppy or a kitten (Shibata and Yukawa 2014, 33). According to Shibata, aside from personal likes and dislikes, puppy and kitten robots raise too many expectations before the interaction, and they are likely to receive criticism with regard to their size, how they feel to the touch, and in terms of what they can do (Shibata and Yukawa 2014, 33). With a seal robot, although a user's initial assessment of it is usually not so high, the more the user interacts with it, the more positive feedback he or she has about it (Shibata and Yukawa 2014, 33). Because a harp seal is not a familiar animal in Japan, users tend not to compare it to a real animal (Masahiro Mori 2014, 183). Reminiscent of a human baby, Paro has a rather large head with two large, black eyes. Its size is about 55 cm in length and its weight is 2.5 kg (at the ninth generation of production) (Shibata and Yukawa 2014, 33). The robot seal is covered with artificial white fur with an antibacterial coating, as it is designed to be touched, caressed, and cuddled. Paro is recharged with a specially designed electric plug that looks like a baby's pacifier. In terms of the visceral level of Norman's schema for design, Paro embodies cuddliness and cuteness.

Paro's behavioural level of appeal is also well calculated. Paro is equipped with sensors for light, sound, touch, temperature, balance; actuators (to generate movement) for its head, eyelids, front flippers, and tail flipper; and a microphone and speakers (Wada and Shibata 2007, 973). Through a combination of movement and a crying sound based on that of a baby seal, the robot seal simulates a 'real' baby seal. Paro responds to touch, light, and sound stimuli. It wiggles slowly with a squeaking sound when being touched. Its main behavioural design feature is its emulation of the helplessness and cuteness of baby animals in general. It signals happiness with facial and gestural means when caressed or hugged. As well, it shows signs of unhappiness when treated badly, for example, if it is hit, left for long periods unattended, or dropped. It is programmed to be active in the daytime and inactive at night-time. For the domestic market, Paro can recognise a few words such as 'hello' or 'cute' in Japanese (or in non-Japanese languages for export models) (Shibata and Yukawa 2014, 33). It can remember its 'own name' and learn to respond to it. Paro also adjusts its behaviour in the course of interacting with its owner, what Breazeal describes as 'socially receptive' in her discussion of socially capable humanoid robot, as it can 'learn from interacting with humans' (2002, 169).

It is obvious that as a socially assistive robotic device for cognitively impaired elderly patients, Paro is designed to induce the visceral and behavioural levels of engagement for its interactants in aged-care facilities in a much more direct and robust way than AIBO's design allows. Paro's reflective level of interaction with the elderly has been explained in terms of its mediating function for the patient's sociality. According to Shibata, for example, Paro can help the patient to recall a memory of an old pet associated with their childhood, which can generate conversations with care givers or other patients (Shibata and Yukawa 2014, 35). Paro functions as a medium to increase opportunities for patients' social interaction.

Shibata states that compared with AIBO, Paro is much more successful in alleviating patients' loneliness (2015, 19).

However, the use of socially assistive robots in aged care, as in the case of Paro, is not ethically neutral. One of the notable concepts shaping the development of these robots in Japan is '*hikikomi genshō*' (the phenomenon of entrainment) (Watanabe 2003; Ogasawara et al. 2004; Nishida 2005; S. Mori et al. 2014). A biological concept, 'entrainment' refers to the alignment of an organism's circadian rhythm with another's rhythm in a shared environment after a certain time spent together. Thus, Japanese robot researchers argue, with reference to the concept of entrainment, that it is important to generate a situation where a human and a robot can maintain their embodied interaction through 'mutual adaptation' (Ishiguro and Washida 2011, 121). This 'interaction', however, is not equal but is based upon an implicit power relation. The noun '*hikikomi*' relates to '*dōka*' (assimilation or adaptation) and '*dōchō*' (synchronisation, synchrony, sympathy, alignment, or conformity). What these concepts point toward is the synchronisation of the user and the robot, suggesting that robot design should take advantage of the fact that human users are capable of adapting themselves to objects in the everyday environment. The designing and theorisation of the communication robot incorporate an expectation and anticipation of the users' adaption to and alignment with robots but not the other way around.

The use of the robot in elderly care, in essence, operates on the same principle, expecting the patients to work with and accommodate the robots. For example, while aged-care patients would not be forced to participate in sessions with robots, they are 'encouraged' by care facilitators to participate in the robot-assisted activity, unlike the owners of AIBO who make their own decisions to purchase and play with the robotic machine. Staff members are actively performing the role of game-show host, creating an appropriate environment for robot performance, the *mise en scène* for the effective use of those robots: for example, using socially assistive robots for games and physical exercise, putting robots in costumes to establish them as 'characters', or giving patients gadgets to play with, such as the use of remote controllers, like the Wii Board, and so on. In some experiments, remote-controlling of a socially assistive robot is actively used to see how a patient would react to a robot that is more 'expressive' (Kimura 2012, 637).

The inviting and participative quality of *matsuri* (non-everyday gatherings, as I have explained in earlier chapters) is operative at robot-assisted activities. In other words, as for the *matsuri*, participants are encouraged to conform to the rules of the event. In turn, the robot-assisted activities work to temporarily modify the 'strangeness' of elderly patients' behaviour by improving their sociality and responsiveness to routine and rules.

Critics of social robotics articulate concerns that social robots enforce normalisation in everyday contexts. As Turkle has similarly observed, Rodogno states that there is a fear that a 'wide-spread diffusion' of social robots will gradually and permanently 'replace human or pet companionship as the standard form of companionship' (2015, 9). Media studies scholar Timo Kaerlein also elaborates upon the issue of 'standardisation and conventionalization' in social robotics, discussing it as facilitating 'reductions of complexity' and controlling the flow of information (2015, 77 and 84). While he recognises that simplified ways of communicating are often necessary and important in developing socially assistive robots, Kaerlein warns of the unintended consequences of such simplifications, especially where these machines interact with people with 'a diminished ability to correctly interpret complex human behavior' (2015, 81). For Kaerlein, social robotics research can be understood as 'a contemporary mode of governance and regulation that privileges certain kinds of subjectification via communication' (2015, 79). In that process, diversity is discouraged and 'the option of nonparticipation – to abstain from communication altogether' is not permitted (Kaerlein 2015, 86). Social robotics would potentially regulate social behaviours, seeing them as 'programmed and reprogrammable' through modelling (Kaerlein 2015, 84).[16] Such concerns become particularly acute in institutional situations, where the question of management and control or, at the very least, that of surveillance, are paramount.

The stated purpose for the use of AIBO and Paro in aged care is as a medium of interpersonal exchange that assists with socialisation. However, elderly patients with cognitive impairments may not be in a position to opt out of assistive activity sessions with facilitating staff or to question their involvement in them; further, the very institutional context itself enforces a normalisation and standardisation that is aligned with amenable and compliant behaviours.

THE INSTITUTIONALISED OTHER

Aiming to address the caregiving needs of a rapidly ageing Japanese population, the use of socially assistive robots for aged care is in fact an experiment that is based on the idea of 'technological fix', as part of Japan's greater reliance on technology for twenty-first-century solutions to its economic and social problems. Research projects conducted within aged-care facilities, already liminal spaces within contemporary Japanese society, are perceived as valid and appropriate. Folklorist Norio Akasaka discusses the tradition of the scapegoat in medieval Japan, in which the old and disabled were forced to live at the margins of society (1992, 260). Aged-care facilities are one such contemporary site, and its cognitively impaired residents, such as those living with forms of dementia, are doubly marginalised. Importantly, the exchange between these robots and elderly patients is meant to be kept within the boundaries of care institutions. There is no openness to the 'outside' in the involvement of patients with social robots: it is an entirely controlled interaction staged in a hermetic environment, and there is no aperture for external perspectives.

Conversely, what goes on in the institution is merely a hothouse version of the interactions that socially assistive robots are designed to cultivate, as I suggested earlier in discussion of AIBO. Roboticist Michio Okada tells the story of witnessing an old woman seated alone in a public park, talking to a small social robot. Watching her talking to her robot, he experienced guilt, pity, and a sense of embarrassment, as well as, contradictorily, a positive attitude that he was witnessing a (normal) episode in a contemporary, technologised society (Okada and Kotaro Matsumoto 2014, i–ii).[17] Importantly, the woman was not a dementia patient. Okada notices that the old woman's talking compensated for the robot's technological deficiencies. Okada's guilt was occasioned by the technological inadequacy of the robot, which he felt was too obvious: as a roboticist, he was disappointed and felt bad for the woman (Okada and Kotaro Matsumoto 2014, ii–iii). The main reason for the shock and embarrassment was that, he states, the incident took place at a public park, a location outside of a care facility. As this incident is detailed in the introduction to his book on the ethics of social robotics, Okada surely recognises that social robots encourage precisely the confusion between real and imagined that he witnessed and, more pertinently, the conflation of real relationships with those that must be created and sustained by the human participant. The old woman's talking alone to a social robot in public, as if it were a pet or a child, reveals both the underside of the new everyday normality that roboticists wish to see develop—and the loneliness of modern life that they perhaps also wish to ameliorate with the solution of companion robots. For Okada, the incident represented an example of the Japanese social phenomenon *muen shakai* (literally meaning 'no relationship society'), referring to an alienated society characterised by broken personal and family relationships and populated by isolated individuals (Okada and Kotaro Matsumoto 2014, ii).[18]

While the reader of Okada's narrative does not know the situation of the elderly woman in the park talking to her robot, it is also possible that she was a robot enthusiast, one of a number of middle-aged to elderly fans of small social robots, like AIBO. I will close the chapter with a brief discussion of these fans. One might not think that a fan culture would also thrive where the use of robots is promoted from the top down, as it has been for social robots in aged-care institutions.

FANDOM: LIVING WITH ROBOTS

Enthusiastic Japanese AIBO owners/fans, generally a demographic that is middle-aged and older, can be very serious about their affection for their robot dogs. They are already crossing the line between real and imagination. Sony stopped its production of AIBO in 2006. It also closed its customer service offering for AIBO maintenance and repair in 2014. Nevertheless, many AIBO owners want to keep their robots operational through yearly maintenance. AIBO's physical movements are often unpredictable, and, if left alone, the robot might fall from a step or attempt to keep moving even after it gets stuck in the furniture, for example (A. Kubo 2015, 216). Writer Mizore Matsumoto comments that owners may also want their AIBO to stay 'alive' because of their memories of dead partners or their appreciation for their AIBO, which they feel gave them 'encouragement' during a difficult period of rehabilitation (2015). There are the elderly AIBO enthusiasts, not dissimilar from the elderly woman in the park described by Okada, like 'Ms. Maekawa, who is 72, talks to the Aibo every day, travels with it and makes clothing for it' (Mochizuki and Pfanner 2015). Nobuyuki Norimatsu, a retired Sony technician, has been offering AIBO repair since 2011 through his company, A-FUN, a company that specialises in repairing vintage electronic products. He has become the last repairman for existing AIBO owners (Mochizuki and Pfanner 2015). Aiming at 'reuniting the robot dogs' spirits with those of their owners', Norimatsu conducted a Buddhist ceremony at a temple in early 2015 for 18 broken AIBO, which he had collected 'from owners who have died' (Mochizuki and Pfanner 2015). Norimatsu uses the functioning parts of these AIBO to repair the robot dogs belonging to other owners.

As I discussed in previous chapters, in Japan, the non-human world can be accommodated conceptually through the traditions of animism (spirits inhering in things) and Buddhism (a dualistic yet complementary under-standing of the self in relation to the external world). Both traditions promote the idea that humans and non-humans are viewed intrinsically as connected. The robot's radical difference is perceived through what might be termed a functional anthropomorphism; at the same time, by seeing them in such a way, human interactants can develop a certain affection for the non-human and, in this case, for robots. AIBO owners are attached to their robots and often sustain a belief in the singular uniqueness of their machines.

The themes of the uniqueness of a communication robot and the recycling of parts of non-functional machines has been explored in robot manga/anime. In essence, AIBO represents an actualisation of similar themes already depicted in the manga fiction of *KV-201XR* (1990), and *Okaimono* (shopping) (1998), a year before AIBO's debut. In the former manga story, an old man lives with a female android that resembles his dead wife when she was young. The old man needs to find the battery pack for the gynoid due to the discontinuation of the product. In the latter work, a domestic service android with the appearance of a young female asks its elderly male owner for permission to be upgraded, even though an improve-ment will delete its memory and replace its body. The fate of out-of-date machines in a capitalist system and the continuation of machine 'life' are paralleled in AIBO repair stories. It is worth observing, though, that it is female robots that require an 'upgrade' for the benefit of their (mostly male) owners. The idea of a new synergy between humans and machines in manga/anime has been explored in dramas featuring domestic care robots of young female appearance that live with elderly men.[19] Such an encounter, even though it is conceived in terms of an imagined future, prepares readers for the idea of cohabitation with robots. Such narratives are not that far from the loving fandom of social robots like AIBO and recall *otaku* sexuality. In Turkle's words, '[i]n Japan, enthusiasm for robots is uninhibited. Philosophically, the ground has been prepared' (2011, 146). While Turkle refers to the influences of Shintoism and Buddhism on the Japanese reception of the robot, I would argue that the cultural influences I have treated in detail in this book, such as manga/anime and their performance-based iterations at expos, or in public squares, or in robot competitions or restaurants, cannot be overlooked.

NOTES

1. Paro was developed by roboticist Takanori Shibata at the National Institute of Advanced Industrial Science and Technology (AIST) in Japan. It was designed as a therapeutic robot to be used at aged-care facilities. Paro has been commercially available since 2004, and was certified by the FDA (the US Food and Drug Administration) as a neurological therapeutic medical device (Class II) for use in the USA in 2009. Paro has been commercially available from Daiwa House Industries since 2010. AIBO (Artificial Intelligent roBOt, which is pronounced in the same way as the Japanese word *aibō*, meaning a pal) was promoted as an interactive robot in a series of autonomous domestic entertainment robots designed and manufactured by Sony. Its first online sale of 5000 units in 1999 sold out in 20 minutes. AIBO represents the epoch of the 'robot boom' in Japan at the beginning of the twenty-first century. Sony discontinued AIBO's production in 2006 due to restructuring. However, in June 2016, Sony announced that it will resume its robotics business.

2. Other types of communication robots, such as NeCoRo (Omron), which has a cat-like appearance, and conversation humanoids such as ifbot (Ifoo) and Palro (Fujisoft), have also been used as 'socially assistive' robots (Yoneoka 2012; Ninomiya 2015).

3. Roboticists Terrence Fong, Illah Nourbakhsh, and Kerstin Dautenhahn distinguish remote-controlled robots from autonomous robots, describing the former in terms of 'peer-to-peer human–robot interaction' and the latter as a 'socially interactive robot' (2003, 145). The more general use of the term social robot in Japan, however, includes tele-operative robotic machines such as Ishiguro's Geminoids. Sociologist Shanyang Zhao's notion of the social robot, on the other hand, includes disembodied computer interaction with humans, such as automated phone response systems or the 'chat-bot' (2006, 406). My discussion of social robots in Japan excludes these disembodied types.

4. The use of robot cleaners for both domestic and professional contexts is steadily increasing. These include Replay by Savioke for hotel use; Pyxis's Helpmate or Panasonic's HOSPI are used in hospitals; and Softbank's Pepper, which has a humanoid appearance and is used as an information service robot.

5. Assistive robotic products developed in Japan include Panasonic's Resyone, a combined device of automated bed and wheelchair; Secom's My Spoon that brings food to a patient's mouth from a dish; Riken's Robear that can carry and transfer a patient between a bed and a wheelchair; and exoskeletal robotic walking devices, such as HAL (Hybrid Assistive Limb) by Cyberdyne, to name a few.

6. These robots include an exercise instructor robot, Bandit (The University of Southern California), and Autom, a weight-loss coach robot (The Massachusetts Institute of Technology); Paro and AIBO for the elderly; for children with developmental disorders, there is Keepon (National Institute of Information and Communications Technology in Japan), a dancing robot, and baby dinosaur robot Pleo (Innvo Labs based in Hong Kong and Nevada, the USA) (Rabbitt et al. 2015); as well as ATR's Robovie, a communication humanoid (Kanda and Ishiguro 2006), and Sony's QRIO, a small and agile dancing humanoid (Tanaka et al. 2007). A team of researchers in Australia, led by health scientist Wendy Moyle, conducted a random controlled trial with Paro (Moyle et al. 2015).

7. The data collected to determine effectiveness includes blood pressure, pulse, temperature, facial expression, amount of activity, patterns of walking movements, and frequency of engagement. The Marte Meo method (an analysis method that involves video recording), Dementia Care Mapping (an observational tool for dementia care), and group interviews with family members and professional carers are also used (Shibata 2012, 8).

8. As of November 2016, the government passed legislation allowing a wider range of foreigners, though numbers remain small, to acquire the qualifications to work as aides in nursing homes to counter a recognised labour shortage (*Jiji Tsūshinsha* [*Jiji News Service*] 2016). In recent years, a limited number of students from Southeast Asian countries have been allowed to train to work as nursing aides in Japan. However, government requirements are stringent and the level of acceptance of such workers in the aged care industry remains to be determined.

9. In Japan, the use of robots in therapeutic contexts is concentrated on aged care because supporting programmes for children with learning difficulties are seen as already sufficiently developed (Yokoyama 2012, 600).

10. At the time of writing, the latest figure for the percentage of people in Japan over the age of 65 is now 26.7 %, as published by the Statistics Bureau (The Statistics Bureau, Ministry of Internal Affairs and Communications 2016b).

11. Recent market research in Japan indicates that assistive robots used for transporting and monitoring elderly patients have been successfully introduced to the institutional market in 2015, and predicts that the Japanese market for assistive robots will be well established by 2020 (Yano Research Institute 2016).

12. The examples include Palmi by Fujisoft and RoBoHon by Sharp and the robot designer Tomotaka Takahashi. These tabletop humanoids can walk, dance, and converse with humans. Palmi can gather news and weather information from the Internet. RoBoHon's multi-functionality includes phoning, data projection, alarm, schedule reminder, reading out incoming email messages, and composing an outgoing email from a voice message. The current buzzword 'The Internet of Things (IoT)' frames these types of robotic devices (The Ministry of Economy, Trade and Industry 2015b, 4, original English).

13. AIBO's presentation has included a stuffed-toy-dog-like appearance, with two eyes, in 2001–2, and a later robot-like model with a dark-coloured face cover, like ASIMO's.

14. Takao Yamaguchi, who was the project manager responsible for the AIBO products at Sony Entertainment Robot Company, expresses reservations regarding Handy Viewer, an accessory product that wirelessly receives 'text messages' from AIBO, as it deviates from the essence of AIBO's visual appeal (2006, 243). Similarly, Yamaguchi states that there were opposing views on AIBO's verbal communication function that later models were equipped with (Yamaguchi 2006, 244).

15. According to Shibata, every Paro is checked with regard to voltage testing, drop testing, and 'hundred thousand times of touching test', and it is equipped with a magnetic shield for a user of a heart pacemaker (2015, 19, original English). Wada and Shibata stress Paro's robustness for long-term use, as compared to other communication robots such as AIBO and NeCoRo (2007, 973).

16. Kaerlein discusses Ishiguro's tele-mediated conversation devices, Telenoid, Elfoid, and Hugvie, as examples toward such modelling and simplification (2015: 78). Ishiguro's android research is indeed

meant to be a means to examine the complexity of human communication and, ultimately, to model it for future robots that will participate in society (Ishiguro and Washida 2011, 141). From a similar perspective, Breazeal argues that these models could be used to understand 'social behavioural disorders' (2002, 1).

17. This story is introduced in the prologue of an edited book by Okada and [Kotaro] Matsumoto, which is used as a 'stepping stone' to discuss ethics regarding social robotics (Okada and Kotaro Matsumoto 2014, iii).

18. The story regarding the old woman in Okada's story can be regarded as doubly disturbing in a society where interpersonal communication is an issue, especially for the younger generations. A decade earlier, Sena warned about the possibility that people would only be able to communicate with social robots, relating to the *sekai kei* works of fiction that depict insular worlds of make protagonists (Sena 2004, 706). (I discussed these works in Chap. 6.) Sena had questioned if the spread of the social robot would further exacerbate the interpersonal communication problems of the generation of Japanese that has already been exposed to the idea of limited human communication through these fictions (Sena 2004, 706–7).

19. The figure of the female carer is often deployed in robot manga/anime works that deal with old-age-related topics. For example, the character Dali in Tezuka's manga *Reunion with Dali* (1982) is, in fact, set up as a female nurse, even though it is not a robot of human appearance. In animation *Roujin Z* (1991), Mr. Takazawa, a bedridden old man, is connected to Z-001, an automated assistive bed equipped with a life-support system, while the voice of his dead wife talks to the old man via the robot bed, which makes him feel like he is being emotionally supported by his wife.

References

Akasaka, N. (1992). *Ijinron josetsu* [Introduction to a study of the outsider]. Tokyo: Chikuma Shobō.

Banks, M. R., Willoughby, L. M., & Banks, W. A. (2008). Animal-assisted therapy and loneliness in nursing homes: Use of robotic versus living dogs. *Journal of the American Medical Directors Association, 9*(3), 173–177. doi:10.1016/j.jamda.2007.11.007.

Bekey, G. A. (2012). Current trends in robotics: Technology and ethics. In P. Lin, K. Abney, & G. A. Bekey (Eds.), *Robot ethics: The ethical and social implications of robotics* (pp. 17–34). Cambridge, MA: MIT Press.

Bender, S. (2012). Robots and the human sciences. *Keisokuto Seigyo, 51*(7), 644–648.

Borenstein, J., & Pearson, Y. (2012). Robotic caregivers: Ethical issues across the human lifespan. In P. Lin, K. Abney, & G. A. Bekey (Eds.), *Robot ethics: The ethical and social implications of robotics* (pp. 251–265). Cambridge, MA: MIT Press.

Breazeal, C. L. (2002). *Designing sociable robots*. Cambridge, MA: MIT Press.

Broekens, J., Heerink, M., & Rosendal, H. (2009). Assistive social robots in elderly care: A review. *Gerontechnology, 8*(2), 94–103. doi:10.4017/gt.2009.08.02.002.00.

Feil-Seifer, D., & Mataric, M. J. (2005). Defining socially assistive robotics. In *Proceedings of the IEEE 9th international conference on rehabilitation robotics* (pp. 465–468). http://ieeexplore.ieee.org/xpl/mostRecentIssue.jsp?punumber=10041#. Accessed on 10-10-2015.

Fong, T., Nourbakhsh, I., & Dautenhahn, K. (2003). A survey of socially interactive robots. *Robotics and Autonomous Systems, 42*(3–4), 143–166. doi:10.1016/S0921-8890(02)00372-X.

Hamada, T., & Sano, T. (2011). Kōreisha serapī yō robotto no inshō ni kansuru chōsa [A study on impressions upon the elderly of therapy robots]. *Tsukubagakuin Daigakukiyō, 6*, 43–48.

Hamada, T., Hashimoto, T., Akazawa, T., & Matsumoto, Y. (2003). Robotto Serapī no Kanōsei ni Kansuru Ichi kōsatsu [A thought on possibilities for robot therapy]. *Kansei Testugaku, 3*, 98–109.

Ishiguro, H., & Washida, K. (2011). *Ikirutte nanyaroka: Kagakusha to tetsugakusha ga kataru, wakamono no tame no kuritikaru jinsei sinkingu* [What does it mean to live: A conversation between a scientist and philosopher, critical life thinking for young people]. Tokyo: Mainichi Shimbunsha.

Jiji Tsūshinsha (Jiji News Service). (2016, November 18). Gaikokujin kaigoshi wo zenmen kaikin: Kanren nihou seiritsu, jisshū mo ukeire [Lifting restrictions on foreign care workers: Two related laws passed; internships also permitted]. *Jiji Dotto Komu*. http://www.jiji.com/jc/article?k=2016111800056&g=pol. Accessed on 11-20-2016.

Kaerlein, T. (2015). Minimizing the human? Functional reductions of complexity in social robotics and their cybernetic heritage. In J. Vincent, S. Taipale, B. Sapio, G. Lugano, & L. Fortunati (Eds.), *Social robots from a human perspective* (pp. 77–88). Cham: Springer International Publishing.

Kameda, T. (2015). Komyunikēshon robotto wo mochiita chiiki fukushi shisetsu deno rekurēshon jisshi no bunseki [A study of the use of communication robots in recreation activities at regional care facilities]. *Sōkajoshi Tankidaigaku Kiyō 46* (February), 9–23.

Kanamori, M., Suzuki, M., Oshiro, H., Tanaka, M., Inoguchi, T., Takasugi, H., Saito, Y., & Yokoyama, T. (2003). Pilot study on improvement of quality of life among elderly using a pet-type robot. In *Proceedings of IEEE international symposium on computational intelligence in robotics and automation* (Vol. 1, 107–112). doi:10.1109/CIRA.2003.1222072.

Kanda, T., & Ishiguro, H. (2006). An approach for a social robot to understand human relationships: Friendship estimation through interaction with robots. *Interaction Studies, 7*(3), 369–403. doi:10.1075/is.7.3.12kan.

Kato, N. (2012). Robotto wo rihabiritēshon ni dōnyū shitemite omou koto [A thought on the introduction of robots to rehabilitation]. *Keisokuto Seigyo, 51*(7), 605–608.

Kimura, R. (2012). Pettogata robotto wo mochi'ite ninchi kōreisha wo taishōnisita robotto serapī [Therapy for the cognitively-impaired elderly with pet-like robots]. *Keisokuto Seigyo, 51*(7), 605–608.

Kubo, A. (2015). *Robotto no jinruigaku: Nijūseiki no kikai to ningen* [Anthropology of the robot: The machine and the human in the 20th century]. Tokyo: Sekai Sisōsha.

Matsumoto, M. (2015, June 1). Petto robo AIBO no shi wo nageki kanashimu kōreisha tachi gijutsusha OB ga oishasan ni nanori wo Ageru [The elderly who suffer grief over the death of their pet robot AIBO, an OB (retired man) technician stands in as a doctor]. *Kyarikone Nyūsu*. https://news.careerconnection.jp/?p=12309. Accessed on 11-20-2015.

Mochizuki, T., & Pfanner, E. (2015, February 11). In Japan, dog owners feel abandoned as Sony stops supporting "Aibo." *Wall Street Journal*. http://www.wsj.com/articles/in-japan-dog-owners-feel-abandoned-as-sony-stops-supporting-aibo-1423609536. Accessed on 11-15-2015.

Mori, M. (2014). *Robotto kōgaku to ningen: Mirai no tameno robotto kōgaku* [Robotics and the human: Robotics for the future]. Tokyo: Ōmusha.

Mori, S., Sano, M., Adachi, N., & Gao, J. (2014). Ningen to robotto no jōdō intarakushon ni okeru tekiō teki hikikomi seigyo to gakushū [Adaptive control and learning of entrainment in affective interaction between humans and robots]. *Intarakushon 2014 Ronbunshū*, February, 232–233.

Moyle, W., Beattie, E., Draper, B., Shum, D., Thalib, L., Jones, C., O'Dwyer, S., & Mervin, C. (2015). Effect of an interactive therapeutic robotic animal on engagement, mood states, agitation and psychotropic drug use in people with dementia: A cluster-randomised controlled trial protocol. *BMJ Open, 5*(8), e009097. doi:10.1136/bmjopen-2015-009097.

NEDO Books henshū iinkai (NEDO Book Editing Committee). (2009). *RT supirittsu: Hito ni yakudatsu robotto gijutsu o kaihatsu suru* [RT spirits:

Developing useful robots for humans]. Kawasaki: New Energy and Industrial Technology Development Organization.

Ninomiya, T. (2015). Komyunikēshon robotto Palro no shōkai to sagami robotto sangyōku ni okeru torikomi [Introducing communication robot Palro and activities at a robot operation area in Sagami]. *Nihon Robotto Gakkaishi, 33*(8), 607–610. doi:10.7210/jrsj.30.1000.

Nishida, T. (2005). Riaru ējento to siteno robotto [A robot as a real agent]. In H. Inoue, T. Kanede, M. Uchiyama, M. Asada, & Y. Anzai (Eds.), *Iwanami kōza robottogaku 5: Robotto infomatikkusu* [Iwanami robot science 5: Robot informatics] (pp. 89–139). Tokyo: Iwanami Shoten.

Norman, D. A. (2004). *Emotional design: Why we love (or hate) everyday things.* New York: Basic Books.

Ogasawara, T., Tajima, T., Hatakeyama, M., & Nishida, T. (2004). Hikikomi genshō ni motozuku ningen to robotto no anmoku jōhō no komyunikēshon [Communication of unconscious information between robots and humans based on the phenomenon of entrainment]. *Jinkōchinō gakkai Zenkoku taikai Ronbunshū* JSAI04: 115–115. doi:10.11517/pjsai.JSAI04.0.115.0.

Okada, M., & Matsumoto, K. [Kotaro] (2014). Prorōgu [Prologue]. In M. Okada & K. [Kotaro] Matsumoto (Eds.), *Robotto no kanashimi: Komyunikēshon wo meguru hito to robotto no seitaigaku* [Sorrow of the robot: Ecology of communication between humans and robots] (pp. i–v). Tokyo: Shinyōsha.

Rabbitt, S. M., Kazdin, A. E., & Scassellati, B. (2015). Integrating socially assistive robotics into mental healthcare interventions: Applications and recommendations for expanded use. *Clinical Psychology Review, 35*, 35–46. doi:10.1016/j.cpr.2014.07.001.

Robertson, J. (2007). Robo sapiens Japanicus: Humanoid robots and the posthuman family. *Critical Asian Studies, 39*(3), 369–398. doi:10.1080/14672710701527378.

Rodogno, R. (2015). Social robots, fiction, and sentimentality. *Ethics and Information Technology,* 1–12. doi:10.1007/s10676-015-9371-z.

Schechner, R. (2013). *Performance studies : An introduction* (3rd ed.). New York: Routledge.

Sena, H. (2001). *Robotto Seiki.* Tokyo: Bungei Shunjū.

Sena, H. (Ed.). (2004). *Robotto Opera (Robot Opera: An anthology of robot fiction and robot culture, original English title).* Tokyo: Kōbunsha.

Sena, H. (2008). *Sena Hideaki robottogaku ronshū* [Hideaki Sena robot study essay collection]. Tokyo: Keisō Shobō.

Sharkey, N., & Sharkey, A. (2012). The rights and wrongs of robot care. In L. Patrick, A. Keith, & G. A. Bekey (Eds.), *Robot ethics: The ethical and social implications of robotics* (pp. 267–282). Cambridge, MA: MIT Press.

Shibata, T. (2010). Ippankatei ni okeru azarashigata robotto Paro tono fureai [Interaction with Paro at Home]. *Robotto, 197*(November), 24–27.

Shibata, T. (2012). Serapī yō robotto Paro no kenkyū kaihatsuto kokunaigai no dōkō [Local and international trends concerning the study and development of therapy robot Paro]. In *Azarashigata robotto Paro niyoru robotto serapī kenkyūkai, shōrokushū* [Therapy robot study group with seal-like Paro, short essay collection] (pp. 4–17). Tokyo: AIST, Shutodaigaku Tokyo, IEEE RAS.

Shibata, T. (2015). Azarashigata robotto niyoru shinkeigaku teki serapī [Neurological therapy with seal-like Paro]. *Seimitsukō Gakkaishi, 81*(1), 18–21.

Shibata, T., & Yukawa, T. (2014). Mentaru komittomento robotto Paro, Fukushi no genba deno katsuyō to iyashi [Mental commitment robot Paro: Use and therapy at a welfare facility]. *Gekkan Fukushi, 97*(8).

Sorell, T., & Draper, H. (2014). Robot carers, ethics, and older people. *Ethics and Information Technology, 16*(3), 183–195. doi:10.1007/s10676-014-9344-7.

Sparrow, R., & Sparrow, L. (2006). In the hands of machines? The future of aged care. *Minds and Machines, 16*(2), 141–161. doi:10.1007/s11023-006-9030-6.

Suzuki, H., Nishi, H., & Taki, K. (2009). Ningen no kansei ni motozuku dōbutsugata robotto no tameno yonkyaku hoyō seisei [Generation method of quadrupedal gait based on human feeling for animal type robot]. *Chinō to Jōhō, 21*(5), 653–662.

Tamura, T., Yonemitsu, S., Itoh, A., Oikawa, D., Kawakami, A., Higashi, Y., Fujimooto, T., & Nakajima, K. (2004). Is an entertainment robot useful in the care of elderly people with severe dementia? *The Journals of Gerontology. Series A, Biological Sciences and Medical Sciences, 59*(1), 83–85.

Tanaka, F., Cicourel, A., & Movellan, J. R. (2007). Socialization between toddlers and robots at an early childhood education center. *Proceedings of the National Academy of Sciences of the United States of America, 104*(46), 17954–17958.

The Ministry of Economy, Trade and Industry. (2015b). New robot strategy: Japan's robot strategy: Vision, strategy, action plan. http://www.meti.go.jp/english/press/2015/pdf/0123_01b.pdf. Accessed on 06-15-2015.

The Statistics Bureau, Ministry of Internal Affairs and Communications. (2013, September 15). Kōreisha no jinkō [The elderly population]. *Tōkei Topikkusu No. 72*. http://www.stat.go.jp/data/topics/topi721.htm. Accessed on 05-20-2015.

The Statistics Bureau, Ministry of Internal Affairs and Communications. (2016a, February 26). Heisei 27 nen kokusei chōsa [The 2015 population census]. http://www.stat.go.jp/data/kokusei/2015/kekka.htm. Accessed on 03-15-2016.

The Statistics Bureau, Ministry of Internal Affairs and Communications. (2016b, July 29). Chūshutsu sokuhō shūkei kekka [The statistics result of the preliminary sample tabulation]. http://www.stat.go.jp/data/kokusei/2015/kekka.htm. Accessed on 06-30-2016.

Turkle, S. (2011). *Alone together: Why we expect more from technology and less from each other*. New York: Basic Books.

Wada, K. (2012). Kōreisha shisetsu ni okeru Paro no unyō hōhō [The method of use for Paro at aged care facilities]. In *Azarashigata robotto Paro ni yoru robotto serapī kenkyūkai, shōrokushū [Therapy robot study Group with seal-like Paro, short essay collection]* (pp. 18–20). Tokyo: AIST, Shutodaigaku Tokyo, IEEE RAS.

Wada, K., & Shibata, T. (2007). Living with seal robots: Its sociopsychological and physiological influences on the elderly at a care house. *IEEE Transactions on Robotics, 23*(5), 972–980. doi:10.1109/TRO.2007.906261.

Watanabe, T. (2003). Shintai komyunikēshon ni okeru hikikomi to shintaisei: Kokoro ga kayou shintaiteki komyunikēshon sisutemu E-Cosmic no kaihatsu wo tōshite [The phenomenon of entrainment and the body in bodily communication: The expression of soul in the development of communication system e-cosmic]. *Bebī Saiensu, 2*, 4–12.

Weiss, A., Wurhofer, D., & Tscheligi, M. (2009). "I love this dog"—Children's emotional attachment to the robotic dog AIBO. *International Journal of Social Robotics, 1*(3), 243–248. doi:10.1007/s12369-009-0024-4.

Yamaguchi, T. (2006). Kateiyō robotto shijō wo kaitakushita AIBO [AIBO, which has opened the home robot market]. In P. H. P. Kenkyūjo (Ed.), *Otonano tameno robotto gaku [Study of the robot for adults]* (pp. 225–251). Tokyo: PHP Kenkyūjo.

Yano Research Institute. (2016). Kaigo robotto shijō ni kansuru chōsa wo jisshi [Implementation of research concerning the market for nursing care robots], Press Release. https://www.yano.co.jp/press/press.php/001546. Accessed on 10-10-2016.

Yokoyama, A. (2012). Rinshō seishinigaku kara mita, robotto serapī no mirai: Animaru serapī no chiken mo kangamite [The future of robot therapy from the perspective of clinical psychiatry: With knowledge of animal therapy]. *Keisokuto Seigyo, 51*(7), 598–602.

Yoneoka, T. (2012). Kōreisha shisetsu deno robotto serapī [Robot therapy at aged care facilities]. *Keisokuto Seigyo, 51*(7), 609–613.

Zhao, S. (2006). Humanoid social robots as a medium of communication. *New Media & Society, 8*(3), 401–419. doi:10.1177/1461444806061951.

CHAPTER 9

Epilogue: Staging a Robot Nation

The idea of the robot as an artificial human being, *jinzō ningen*, was introduced to Japan through the theatre production of Capek's *R.U.R.* in Tokyo in 1924. Nearly one hundred years later, Japan is planning to present to the world its version of a robot-based society in 2020, the year of the second Tokyo Olympics. In early 2015, the Japanese government released the report 'Japan's Robot Strategy', which endorses the idea of a 'New Industrial Revolution Driven by Robots'. It compiles expert discussions organised through the Robot Revolution Realisation Council, which was set up by the Japanese government as part of the Japan Revitalization Strategy of 2014 (The Ministry of Economy, Trade and Industry 2014). Through this council, specialists from diverse backgrounds discussed 'the utilization of robot technologies to improve Japan's productivity, enhance companies' earning power, and raise wages' (The Ministry of Economy, Trade and Industry 2015a, original English). The report focuses on 'Japan as robot innovation hub', 'showcasing [Japanese] robots to the world'; Japan will lead robotics, formulating and disseminating business rules and standards for robot technologies internationally (The Ministry of Economy, Trade and Industry 2015a, original English). The document discusses a five-year plan to achieve these objectives, heading toward 2020, including the idea of the 'Robot Olympic (Provisional Name)' (The Ministry of Economy, Trade and Industry 2015b, 48, original English).[1]

A sense of crisis and urgency permeates the 'Japan's Robot Strategy' report. It argues for the 'robot revolution' through familiar rhetoric, pointing out Japan's ageing population and its labour shortages and citing

Y. Sone, *Japanese Robot Culture*,
DOI 10.1057/978-1-137-52527-7_9

the country's current position economically and politically in relation to recent developments in global robotics. Although Japan still maintains 'its status as "Robotics Superpower"'—it is 'the world's number one supplier of industrial robots in value', and it is also at the top 'in the number of units in operation till present'—Japan's weakening competitiveness in the manufacturing sector has been acknowledged (The Ministry of Economy, Trade and Industry 2015b, 2, original English). Countries now recognise the growth potential in robotics and pour funds into the R&D of this field through government-led projects and initiatives. These projects include the US government's 'National Robotics Initiative' in 2011; Europe's 'EU SPARC Project' in 2014; and China's 'Development Plan for Intelligent Manufacturing Equipment Industry' in 2012. Google's acquisition of seven robot venture companies in 2014 to gain cutting-edge robot technologies is also discussed in the report (The Ministry of Economy, Trade and Industry 2015b, 3, original English). With reference to the tough economic situation, the report stresses the need to explore product areas that have not yet been automated and robotised, such as in the car, electronics in the home, and developments in mobile phone technology. The language of the report is assertive: it argues that Japan must successfully achieve its robot revolution before other countries do and work on 'establishing a platform status to win global competitions' (The Ministry of Economy, Trade and Industry 2015b, 16, original English). The report's authors regard 'entire Japan' as the site for demonstrating and implementing new and innovative robot technologies and for the idea of human-robot cohabitation, 'bringing about a "daily life with robots" in society', not only in industry but in everyday life (The Ministry of Economy, Trade and Industry 2015b, 10, original English). The report argues for the importance of integrating interactive, next-generation social robots in domestic and institutional settings.

In its techno-determinist economic rationalism, the government report does not indicate why such integration of robots in Japanese society would bring significant social benefits. There is an iterative relationship between claims that it is necessary to implement such policies concerning robots and that these robots are socioeconomically beneficial. It seems that policy makers have devised the solution first, the development of next-generation social robots, and now seeks 'problems' to solve, settling fixedly upon Japan's population of elderly and its decreasing working population. The complexities of existing social circumstances are now called upon to fit the desired solution. The tone of the report gives the impression that people are the 'test subjects' for data gathering as well as 'human exhibits' for the occasion of the Olympics, as a large-scale *matsuri*, set to perform for the foreign media and visitors. The underlying forces behind such government directives can be discussed in terms of what Shunya Yoshimi calls '*omatsuri shugi*' (*matsuri*-ism with the honorific 'o', used ironically here) (2015). It is a disparaging phrase to highlight the nationalist, ideological hunger for large-scale, government-sanctioned festival events. This would not be the first time that Japan has seen *o-matsuri*-ism: the entire nation was urgently exhorted and rallied before and during the Tokyo Olympics in 1964, and the 1970 Exposition in Osaka.

The idea of the 'Robot Olympics' can, of course, be understood as one part of the larger Cultural Olympiad, which is a series of cultural activities required of the city hosting the summer Olympic Games. In this regard, it is inevitable that politics will play a role. In relation to the bids for the Olympic Games (the failed bid for 2016 and the successful bid for the 2020 Games), the Tokyo Metropolitan Government developed cultural policies and implemented pilot projects including the Tokyo Culture Creation Project (2008), Tokyo Art Point Project (2009), and the Networking Project (2011). It established Arts Council Tokyo in 2012, whose task is to 'carry out major cultural projects to prepare for the 2020 Olympic and Paralympic Games in Tokyo' (Arts Council Tokyo 2016, original English). Leading toward the 2020 Games, Prime Minister Shinzo Abe met with a group of selected intellectuals and cultural elites to discuss cultural diplomacy and soft power, in conjunction with the Agency for Cultural Affairs and the Ministry of Foreign Affairs. They tabled the idea of organising a 'Japan Expo', which would promote Japan and Japanese culture, in major global cities during 2018 to commemorate the 150th anniversary of the Meiji Restoration (*Jiji Tsūshinsha* [*Jiji News Service*] 2015).[2] Naturally, local and municipal governments and relevant organisations are also planning events for foreign tourists before and during the Games, to help revitalise their economies.[3]

Many robot-related events, including those with popular themes held by non-government organisations, are likely to be integrated, directly and indirectly, into the promotional regime for the Olympics. For example, in 2014, the Gundam Global Challenge set up an open call for engineering ideas that would enable the 18m high and otherwise static Gundam statue in Odaiba to 'walk' in 2019. This is conceived as a celebratory event for the fortieth anniversary of *Mobile Suit Gundam*, the 1979 original TV series. For the technical director of the project, Shuji Hashimoto, an applied physics expert, the aim of the event would be to generate synergies between dreams and reality (Gundam Global Challenge 2014). For Yasuo Miyakawa, the executive producer of the Gundam Global Challenge project and the president of Sunrise, the company that produces the *Gundam* animation series, the event aims to entertain and touch spectators' hearts (Gundam Global Challenge 2014). Ollie Barder, *Forbes*' Tokyo correspondent, is hoping that 'if they [the Gundam Global Challenge] manage to pull it off then … it will be included in the opening ceremony for the Tokyo Olympics in 2020' (2015, my ellipsis). It is very revealing that even non-Japanese people such as Barder sees the Gundam event in relation to the Olympics. Barder's comments are a symptom of international expectations held by both the Japanese and non-Japanese public, as well as manga/anime fans concerning the centrality of the robot on the mainstream cultural stage.[4]

Masahiko Inami, research engineer in the fields of robotics and augmented reality, is planning to take advantage of the Olympics as *matsuri*. Inami advocates the concept of 'superhuman sports', a new kind of physical games, which meshes human bodies with visual and wearable technologies (Inami 2015).[5] This is not a government-sponsored initiative, though Inami is seeking funding. One of the possible ideas for 'superhuman sports' is the use of Skeletonics Suit, a 2.6-meter high exoskeleton developed by Reyes Tatsuru Shiroku, a key member of the 2008 Robocon champion team. Another idea of Inami's includes the use of drones and projected images onto players' 3D goggles: it is imagined to be a virtual sports event, a sort of virtual 'quidditch', as in the *Harry Potter* series. Developing from his research interest in the notion of the 'augmented human', Inami sees the 2020 Olympics as an opportunity for presenting a uniquely Japanese event that combines sports, technology, and culture (Inami 2015, 59).

While numerous creative people from non-government sectors, such as Inami, seek to participate in the national events of the Olympics in 2020 through art and cultural means, and the themes of robots or use of robotic device are often part of their discussions, it is worth noting that even artists and writers involved in Japan's various subcultures are getting involved, though their aims are decidedly not nationalistic. Tsunehiro Uno of the editor of the journal *Planets* published an issue with the title 'Tokyo 2020 Alternative Olympic Project' (Uno 2015). Uno hopes that the occasion of the Olympics can open an opportunity for the subcultural imagination to reveal the political rhetoric surrounding the Games (2015, 268). Similarly, Hiroyuki Ito, the producer of Crypton Future Media and creator of

Hatsune Miku, holds the view that the *omatsuri* of the Olympics can enliven subcultural expression (in a discussion forum in Uno 2015, 187).

The dual investments of the government and that of non-government sectors in the 2020 Tokyo Olympics epitomise this book's key theme: the dynamic relation of technology, politics, and culture, including popular-culture expression, around the topic of robots in Japan. In Karel Čapek's original play *R.U.R.*, the figure of the robot was seen as an emblem of modern Western technology in Japan in the early twentieth century. Astro Boy became a symbol of Japan's post-war recovery, representing the nation's presumed pacifist approach to advanced technologies. The robotisation and automatisation of factories are seen to be actualising the *Astro Boy* ideology of a bright, technologised future, while robot performances and demonstrations since Expo '70 in Osaka were vivid representations of that vision. The next-generation humanoid robots of interactivity and cohabitation are feverishly promoted by the government and by the robotics industries as signs of a new optimism, pulling Japan out of its economic doldrums in a successful new millennium. Responding to techno-politics, artists and creators in the arts and in subculture contexts continue to express their own views of robots, and many of these views, even those of *otaku* culture, are then absorbed into the Japanese cultural mainstream and promoted through Japan's highly efficient capitalist (and nationalist) organs.

In this book, I have endeavoured to dispense with the generalities of the presumed Japanese affinity for robots, an explanation that all too easily aligns with the cultural-essentialist *nihonjinron* narrative. Through an examination of diverse examples of robot 'performance', I have aimed to highlight particular instances of transformative engagements by specific audiences, in specific situations. These examples have been discussed in terms of their spectacular natures and their creation of particular and culturally contextualised effects and affects. I have articulated the variously familiar, strange, and wondrous agencies granted to these performing robots in social contexts, and I have explained in what ways robots in Japanese culture remain potent cultural ciphers.

For the Japanese public, robot imagery across media, invariably drawn from manga/anime, tends toward a positive perception of the robot; this is highly convenient for the next-generation robot proponents, who readily forge links with cultural traditions such as animism and Japanese pride in its history, such as the development of the *karakuri ningyō*. In Chap. 3, I discussed how such views are absorbed into nationalistic narratives at large-scaled national and international events. Chap. 4 compared mainstream theatre works that similarly perpetuate established government and robot-industry views, while more radical art-making explores precisely the uncanniness of the robot's presence that more mainstream artists wish to ignore. Chap. 5 discussed aspects of 'place': how iconic robots create local and corporate meaning for the Japanese, while Fantasy Japan, which is at the same time Edenic, erotic, and exotic, is staged for foreigners at Robot Restaurant. As the chapters progress, one's physical proximity to my robot subjects becomes closer and closer. Chap. 6 explores *otaku* fandom and sexual affinity for Hatsune Miku, while Chap. 7 looks at robot competitions, exploring the transformative potential of the game for all of its participants. Finally, in Chap. 8, I consider how patients in aged care facilities are managed (and experimented upon) to like socially assistive 'pet' robots, an engagement achieved through careful contextual framing and facilitation.

The robot remains a lens for the Japanese of ideology and tradition. This book has examined how specific modes of encounter between the audience (or users) and robots in sensually and narratively heightened *mises en scène*—against the backdrop of the question of Japan's modernity, always set to counter the superior powers of the West—can trigger reflexive and culturally specific response. I have made connections across public and private forms of the Japanese engagement with robots, which have not been examined in relation to each other. The transformative power of these robot performances is unique to each of the *mises en scène* that situates the performing robot in ways that are precisely calibrated to the investments of the Japanese government or corporate concerns, the latter incorporating the needs and desires of hobbyists and fans. Those hobbyists and fans are influenced by the dynamics of their subcultures, and they, in turn, influence mainstream culture: witness the formidable soft-power presence of *otaku* culture and the popularity of manga/anime worldwide. The robot is not uniquely Japanese, but in its uniquely Japanese expressions, it figures forth the Japanese nation's aspirations, desires, and fears more powerfully than does any other cultural icon. Its rhetorical and metaphoric power for the Japanese, its generativity as a symbol, emerges from an iterative loop between the real and representation in Japanese thought: culture is constituted by the very expressions that are supposed be its product. The robot in Japan is both performing and performed, just as the Japanese public is expected, always and unfailingly, to participate in its dynamic performance.

Notes

1. While the term 'Robot Olympic' in the 2015 report has become widely used, it is likely to end up with a different name due to copyright concerns; currently, a more neutral term, 'International robot competition', is used at the government level (M. Otsuka 2016). The possible games that would feature in the competition would not be based on sports, such as humanoid soccer, as in RoboCup, but would involve practical and pragmatic activities performed by robots designed to be used at factories, in hazardous situations, and in domestic contexts.

2. The Japan Expo has existed in Paris as an annual festival to celebrate Japanese culture—'from manga to fashion, from animation to traditions and music and video games'—since 2000 (Japan Expo 2014). While there is no specific reference to the Paris fair, it is possible to assume that the government is thinking of replicating similar events in other major cities around the world.

3. In 2015, the Kinosaki Onsen, a hot-springs resort in Toyo'oka, located in Hyogo Prefecture in the Kansai region, introduced interactive humanoid Pepper to provide tourist information, and, in the same year, the Hen Na Hotel in Sasebo, Nagasaki, located on Kyūshū Island, opened with robots as part of its staff, attracting foreign media attention. It is foreseeable that an increasing number of gimmicky tourist attractions like these, especially in regional Japan, will be introduced in the lead-up to the 2020 Tokyo Olympics.

4. The centrality of Japanese manga/anime in Japanese cultural politics was evidenced by the fact that Shinzo Abe, as the prime minister of the host country for the 2020 Olympics, appeared as the character Super Mario, along with Japanese popular culture icons such as Hello Kitty and Doraemon, which appeared on video, at the closing ceremony of the Rio Olympics in August 2016.

5. Inami conceives the concept of superhuman sports in relation to the Cybathlon, an international competition for disabled athletes with advanced assistive technology such as powered exoskeletons, robotic prostheses, and brain-computer interfaces. The inaugural Cybathlon, envisioned as an Olympics for 'bionic athletes', was held in Switzerland in October 2016 (Tufnell 2014). Inami and his students participated in the 2008 Bacarobo contest, and won the prize.

REFERENCES

Arts Council Tokyo. (2016). Who we are: About. *Arts Council Tokyo.* https://www.artscouncil-tokyo.jp/en/who-we-are/about/. Accessed on 03-10-2016.

Barder, O. (2015, November 4). Gundam global challenge winners determined to make a full size Gundam walk. *Forbes.* http://www.forbes.com/sites/olliebarder/2015/11/04/gundam-global-challenge-winners-determined-to-make-a-full-size-gundam-walk/. Accessed on 01-15-2016.

Gundam Global Challenge. (2014, July 9). Talk about Gundam Global Challenge! A dream makes reality, and likewise reality makes a dream. *GGC Special Vol. 1.* http://gundam-challenge.com/en/. Accessed on 10-10-2015.

Inami, M. (2015). 2020 nen wa chōjin orinpikkue no tsūkaten ni suginai [The Year 2020 is Merely an Episode in the Journey Toward a Superhuman Olympics]. Tsunehiro Uno (Ed.). *Planets, 9* (January): 58–63.

Japan Expo. (2014). The history of Japan Expo. http://www.japan-expo-paris.com/en/menu_info/history_475.htm. Accessed on 10-10-2015.

Jiji Tsūshinsha (*Jiji News Service*). (2015, December 18). 18 nen no nihon haku kentō: Abe shushō, bunka no miryoku hasshin, seifu kondankai [Considering the Japan Expo. 2018: Prime Minister Abe and the transmission of cultural power at a government round-table conference]. *Jiji Dotto Komu.* http://www.jiji.com/jc/zc?k=201512/2015121800946. Accessed on 01-15-2016.

Otsuka, M. (2016, February 24). 2020 nen no robotto orinpikku wa dōnaru?: Hatsuno shimonkaigi ga kaisai [What could happen to the 2020 Olympics: The first advisory committee meeting was held]. *Mainabi Nyūsu.* http://news.mynavi.jp/articles/2016/02/04/robot_olympic/. Accessed on 03-03-2016.

The Ministry of Economy, Trade and Industry. (2014, June 24). Japan revitalization strategy (Revised in 2014)-Japan's challenge for the future. https://www.kantei.go.jp/jp/singi/keizaisaisei/pdf/honbun2JP.pdf. Accessed on 12-20-2015.

The Ministry of Economy, Trade and Industry. (2015a, January 23). Japan's robot strategy was compiled- action plan toward a new industrial revolution driven by robots. http://www.meti.go.jp/english/press/2015/0123_01.html. Accessed on 06-15-2015.

The Ministry of Economy, Trade and Industry. (2015b). New robot strategy: Japan's robot strategy: Vision, strategy, action plan. http://www.meti.go.jp/english/press/2015/pdf/0123_01b.pdf. Accessed on 06-15-2015.

Tufnell, N. (2014, March 27). Cybathlon 2016: First 'olympics' for bionic athletes. *Wired UK.* http://www.wired.co.uk/news/archive/2014-03/27/cybathlon. Accessed on 01-20-2016.

Uno, T. (Ed.). (2015). *Planets: Tokyo 2020 alternative Olympic project* (Vol. 9). Tokyo: Dainiji Wakusei Kaihatsu Iinkai/Planets.

Yoshimi, S. (2015). Posuto sensō to shite no orinpikku: 1964 nen Tokyo taikai wo saikō suru [The Olympics as an alternative to war: A reexamination of the 1964 Tokyo Olympics]. *Masu Komyunikēshon Kenkyū, 86*(January), 19–37.

REFERENCES

Akagi, A. (1992). Bishōjo shōkō gun, Rorikon toiu yokubō [The beautiful young girl syndrome, the desire called Lolicon]. *New Feminism Review, 3*, 230–234.

Akasaka, N. (1992). *Ijinron josetsu* [Introduction to a study of the outsider]. Tokyo: Chikuma Shobō.

Allison, A. (2006). *Millennial monsters Japanese toys and the global imagination.* Berkeley: University of California Press.

Aoyagi, H. (2005). *Islands of eight million smiles: Idol performance and symbolic production in contemporary Japan.* Cambridge, MA: Harvard University Asia Center.

Aramata, H. (1996). *Daitōa kagaku kitan* [Mysterious science stories in Greater Asia]. Tokyo: Chikuma Shobō.

Arjomand, M., & Mosse, R. (2014). Editors' preface. In M. Arjomand & R. Mosse (Eds.), *The Routledge introduction to theatre and performance studies* (M. Arjomand, Trans.) (pp. viii–ix). London/New York: Routledge.

Arkenberg, C. (2014, May 24). Domo Arigato Restaurant Roboto!. *Boing Boing.* http://boingboing.net/2014/05/24/robot.html. Accessed on 12-10-2014.

Arts Council Tokyo. (2016). Who we are: About. *Arts Council Tokyo.* https://www.artscouncil-tokyo.jp/en/who-we-are/about/. Accessed on 03-10-2016.

Asada, M. (2010). *Robotto to iu sisō* [A philosophy called the robot]. Tokyo: NHK Shuppan.

Asada, M. (2011). Robot to saiensu ga michibiku ugoki, katachi to shikō no aratana kagaku (A new area of scientific research concerned with movement, form, and thought, led by robot science). *Bessatu Nikkei Saiensu, 179*, 4–6.

Asada, M., & Higaki, T. (2013). Robotto, ningen, seimei [Robot, human being, life]. In T. Higaki (Ed.), *Robotto, sintai, tekunolojī: baiosaiensu no jidai ni okeru*

© The Author(s) 2017
Y. Sone, *Japanese Robot Culture,*
DOI 10.1057/978-1-137-52527-7

ningen no mirai [The robot, the body, technology: The future of the human being in the age of bioscience] (pp. 3–35). Osaka: Osaka Daigaku Shuppankai.

Asimov, I. (2004). *I, Robot* (Bantam hardcoverth ed.). New York: Bantam Dell.

Auslander, P. (2006). Humanoid boogie: Reflections on robotic performance. In D. Krasner & D. Z. Saltz (Eds.), *Staging philosophy: Intersections of theater, performance, and philosophy* (pp. 87–103). Ann Arbor: University of Michigan Press.

Azuma, H. (2001). *Dōbutsuka suru posutomodan: Otaku kara mita nihonshakai* [Animalizing postmodern: Japanese society as seen from *otaku*]. Tokyo: Kōdansha.

Azuma, H. (2007). *Gēmu teki riarizumu no tanjō: Dōbutsuka suru posutomodan 2* [The birth of game realism: Animalising the Postmodern 2]. Tokyo: Kōdansha Gendai Shinsho.

Azusa, M. (2009, November 2). Dai jū rokkai ROBO-ONE in Toyama repōt [A report on The 16th ROBO-ONE in Toyama]. *Robot Watch*. http://robot. watch.impress.co.jp/docs/news/326137.html. Accessed on 01-15-2010.

Baba, N. (2004). Honsho no nerai [The objective of this book]. In *Robotto no bunkashi: kikai o meguru sōzōryoku* [Cultural analysis on the robot: The imagination concerning machines] (pp. 7–11). Tokyo: Shinwasha.

Balme, C. B. (2014). *The theatrical public sphere*. Cambridge: Cambridge University Press.

Banks, M. R., Willoughby, L. M., & Banks, W. A. (2008). Animal-assisted therapy and loneliness in nursing homes: Use of robotic versus living dogs. *Journal of the American Medical Directors Association, 9*(3), 173–177. doi:10.1016/j.jamda. 2007.11.007.

Bar-Cohen, Y., & Hanson, D. (2009). *The coming robot revolution: Expectations and fears about emerging intelligent, humanlike machines*. New York: Springer.

Barder, O. (2015, November 4). Gundam global challenge winners determined to make a full size Gundam walk. *Forbes*. http://www.forbes.com/sites/ olliebarder/2015/11/04/gundam-global-challenge-winners-determined-to-make-a-full-size-gundam-walk/. Accessed on 01-15-2016.

Barthes, R. (1993). The world of wrestling. In S. Sontag (Ed.), *A Barthes reader* (pp. 18–30). London: Vintage.

Bartneck, C., Suzuki, T., Kanda, T., & Nomura, T. (2006). The influence of people's culture and prior experiences with Aibo on their attitude towards robots. *AI & Society, 21*(1–2), 217–230. doi:10.1007/s00146-006-0052-7.

Baudrillard, J. (2005). *The system of objects* (J. Benedict, Trans.) (2nd ed.). London/New York: Verso.

Baudrillard, J. (2009). Subjective discourse or the non-functional system of object. In F. Candlin & R. Guins (Eds.), *The object reader* (pp. 41–63). London/New York: Routledge.

Bekey, G. A. (2012). Current trends in robotics: Technology and ethics. In P. Lin, K. Abney, & G. A. Bekey (Eds.), *Robot ethics: The ethical and social implications of robotics* (pp. 17–34). Cambridge, MA: MIT Press.

Bekey, G., Ambrose, R., Kumar, V., Lavery, D., Sanderson, A., Wilcox, B., Yuh, J., & Zheng, Y. (2008). *Robotics: State of the art and future challenges*. London: Imperial College Press.

Bell, J. (2008). *American puppet modernism: Essays on the material world in performance*. New York: Palgrave Macmillan.

Bell, J. (2014a). Omnipresence and invisibility: Puppets and the textual record. In D. N. Posner, C. Orenstein, & J. Bell (Eds.), *The Routledge companion to puppetry and material performance* (pp. 7–10). London/New York: Routledge.

Bell, J. (2014b). Playing with the eternal uncanny: The persistent life of lifeless objects. In D. N. Posner, C. Orenstein, & J. Bell (Eds.), *The Routledge companion to puppetry and material performance* (pp. 43–52). London/New York: Routledge.

Bender, S. (2012). Robots and the human sciences. *Keisokuto Seigyo, 51*(7), 644–648.

Benford, G., & Malartre, E. (2008). *Beyond human: Living with robots and cyborgs*. New York: Forge Books.

Bennett, S. (1990). *Theatre audiences: A theory of production and reception*. London/New York: Routledge.

Bennett, J. (2010). *Vibrant matter: A political ecology of things*. Durham: Duke University Press.

Berghaus, G. (2005). *Avant-garde performance: Live events and electronic technologies*. Houndmills/Basingstoke/Hampshire/New York: Palgrave Macmillan.

Bergson, H. (1935). *Laughter: An essay on the meaning of the comic* (C. Brereton & F. Rothwell, Trans.). London: Macmillan and Co.

Birch, A., & Tompkins, J. (Eds.). (2012). *Performing site-specific theatre: Politics, place, practice*. Basingstoke: Palgrave Macmillan.

Birringer, J. H. (2008). *Performance, technology, and science*. New York: PAJ Publications.

Black, D. (2012). The virtual idol: Producing and consuming digital femininity. In P. W. Galbraith & J. G. Karlin (Eds.), *Idols and celebrity in Japanese media culture* (pp. 209–228). Houndmills/Basingstoke/Hampshire/New York: Palgrave Macmillan.

Blau, H. (1990). *The audience*. Baltimore: Johns Hopkins University Press.

Boenisch, P. M. (2014). Acts of spectating: The dramaturgy of the audience's experience in contemporary theatre. In K. Trencsényi & B. Cochrane (Eds.), *New dramaturgy: International perspectives on theory and practice* (pp. 225–241). London/New York: Bloomsbury.

Bogost, I. (2011). *How to do things with videogames*. Minneapolis: University of Minnesota Press.

Bolton, C. A. (2002). From wooden cyborgs to celluloid souls: Mechanical bodies in anime and Japanese puppet theater. *Positions: East Asia Cultures Critique, 10*(3), 729–771.

Borenstein, J., & Pearson, Y. (2012). Robotic caregivers: Ethical issues across the human lifespan. In P. Lin, K. Abney, & G. A. Bekey (Eds.), *Robot ethics: The ethical and social implications of robotics* (pp. 251–265). Cambridge, MA: MIT Press.

Borggreen, G. (2006). Ruins of the future: Yanobe Kenji Revisits Expo '70. *Performance Paradigm: Journal of Performance and Contemporary Culture, 2,* 119–131.

Borggreen, G. (2014). 'Robots cannot lie': Performative parasites of robot-human theatre. In *Sociable robots and the future of social relations: Proceedings of Robo-Philosophy 2014* (pp. 157–163). Amsterdam: IOS Press.

Breazeal, C. L. (2002). *Designing sociable robots.* Cambridge, MA: MIT Press.

Broadhurst, S. (2009). *Digital practices: Aesthetic and neuroesthetic approaches to performance and technology.* Houndmills/New York: Palgrave Macmillan.

Broekens, J., Heerink, M., & Rosendal, H. (2009). Assistive social robots in elderly care: A review. *Gerontechnology, 8*(2), 94–103. doi:10.4017/gt.2009.08.02.002.00.

Brown, S. T. (2010). *Tokyo cyberpunk: Posthumanism in Japanese visual culture.* New York: Palgrave Macmillan.

Buruma, I. (1984). *Behind the mask: On sexual demons, sacred mothers, transvestites, gangsters, drifters and other Japanese cultural heroes.* New York: Pantheon Books.

Butler, J. (1988). Performative acts and gender constitution: An essay in phenomenology and feminist theory. *Theatre Journal, 40*(4), 519–531. doi:10.2307/3207893.

Cabinet Office, Government of Japan. (2007). Inobēshon [Innovation] 25. http://www.cao.go.jp/innovation/index.html. Accessed on 10-15-2015.

Calichman, R. (2008). *Overcoming modernity: Cultural identity in wartime Japan.* New York: Columbia University Press.

Carlson, M. (1989). *Places of performance: The semiotics of theatre architecture.* Ithaca: Cornell University Press.

Caro, M. (2008). *Astroboy in Roboland.* HD Video. Paris: Les Films d'Ici.

Carpenter, J., Davis, J. M., Erwin-Stewart, N., Lee, T. R., Bransford, J. D., & Vye, N. (2009). Gender representation and humanoid robots designed for domestic use. *International Journal of Social Robotics, 1*(3), 261–265. doi:10.1007/s12369-009-0016-4.

Casserly, M. (2015, August 20). Giant robots prepare to do battle, as America and Japan go to war. And the US needs YOUR help. *PC advisor.* http://www.pcadvisor.co.uk/news/social-networks/america-vs-japan-in-giant-warrior-robot-battle-help-3619707/. Accessed on 12-10-2015.

Causey, M. (2006). *Theatre and performance in digital culture from simulation to embeddedness.* London/New York: Routledge.

Cavalcanti, G., & Oehrlein, M. (2015). The USA vs Japan giant robot duel. *MegaBots.* http://www.megabots.com/. Accessed on 12-10-2015.

Clarke, L. B., Gough, R., & Watt, D. (2007). Opening remarks on a private collection. *Performance Research, 12*(4), 1–3. doi:10.1080/13528160701822544.

Cohen, J. (1966). *Human robots in myth and science*. London: George Allen & Unwin.

Coleman, S., & Crang, M. (2002). Grounded tourists, travelling theory. In S. Coleman & M. Crang (Eds.), *Tourism: Between place and performance* (pp. 1–17). New York: Berghahn Books.

Condry, I. (2013). *The soul of Anime: Collaborative creativity and Japan's media success story*. Durham: Duke University Press.

Connor, S. (2000). *Dumbstruck: A cultural history of ventriloquism*. New York/Oxford: Oxford University Press.

Cox, R. (2002). Is there a Japanese way of playing? In J. Hendry & M. Raveri (Eds.), *Japan at play: The ludic and logic of power* (pp. 169–185). London/New York: Routledge.

Crypton Future Media. (2016a). Hatsune Miku V4X. http://www.crypton.co.jp/mp/pages/prod/vocaloid/mikuv4x.jsp. Accessed on 10-10-2016.

Crypton Future Media. (2016b). Vocaloid 2, Hatsune Miku. http://www.crypton.co.jp/mp/pages/prod/vocaloid/cv01.jsp. Accessed on 04-15-2016.

Davis, T. C., & Postlewait, T. (2003). Theatricality: An introduction. In T. C. Davis & T. Postlewait (Eds.), *Theatricality* (pp. 1–39). Cambridge/New York: Cambridge University Press.

D'Cruz, G. (2014). 6 things I know about Geminoid F, or what I think about when I think about android theatre. *Australasian Drama Studies, 65*, 272–288.

de Certeau, M. (1984). *The practice of everyday life* (S. Rendall, Trans.). Berkeley: University of California Press.

De Marinis, M. (1987). Dramaturgy of the spectator (P. Dwyer, Trans.). *TDR/The Drama Review, 31*(2): 100–114. doi:10.2307/1145819.

Dery, M. (1996). *Escape velocity: Cyberculture at the end of the century*. New York: Grove Press.

Dixon, S. (2007). *Digital performance: A history of new media in theater, dance, performance art, and installation*. Cambridge, MA: MIT Press.

Doherty, M. (2013, March 2). With I, Worker, Canadian Stage takes on the inevitable robopocalypse. *National Post*. http://news.nationalpost.com/arts/on-stage/with-i-worker-canadian-stage-takes-on-the-inevitable-robopocalypse. Accessed on 10-10-2014.

Doi, T. T. (2012). Inugata robotto AIBO to shin robotto sangyō [Robot Dog AIBO and the New Robot Industry]. *Nihon Robotto Gakkaishi, 30*(10), 1000–1001. doi:10.7210/jrsj.30.1000.

Durkheim, É., & Mauss, M. (1963). *Primitive classification* (R. Needham, Trans.). London: Cohen & West.

Eckersall, P. (2013). *Performativity and event in 1960s Japan: City, body, memory*. Houndmills/Basingstoke/Hampshire/New York: Palgrave Macmillan.

Eckersall, P. (2015). Towards a dramaturgy of robots and object-figures. *TDR/The Drama Review, 59*(3), 123–131. doi:10.1162/DRAM_a_00474.

Endo, M. (2002). *Gandamu, ichinen sensō* [Gundam, One Year's War]. Tokyo: Takarajimasha.

Endo, T. (2013). *Sōsharuka suru ongaku: Chōshu kara asobi e* [Music facilitates sociability: From listening toward playing]. Tokyo: Seidosha.

Engelberger, J. F. (1999). Historical perspective and role in automation. In S. Y. Nof (Ed.), *Handbook of industrial robotics* (2nd ed., pp. 1–10). New York: John Wiley & Sons.

Enomoto, A. (Ed.). (2009). *Otaku no kotoga Omosiroi hodo wakaru Hon* [This book easily allows you to understand what an *otaku* is]. Tokyo: Chūkei shuppan.

Feenberg, A. (2010). *Between reason and experience essays in technology and modernity.* Cambridge, MA: MIT Press.

Feil-Seifer, D., & Mataric, M. J. (2005). Defining socially assistive robotics. In *Proceedings of the IEEE 9th international conference on rehabilitation robotics* (pp. 465–468). http://ieeexplore.ieee.org/xpl/mostRecentIssue.jsp?punumber=10041#. Accessed on 10-10-2015.

Fensham, R. (2009). *To watch theatre: Essays on genre and corporeality.* Bruxelles/New York: P.I.E. Peter Lang.

Fischer-Lichte, E. (2008). *The transformative power of performance: A new aesthetics* (S. I. Jain, Trans.). New York: Routledge.

Fischer-Lichte, E. (2014). *The Routledge introduction to theatre and performance studies.* M. Arjomand & R. Mosse (Eds.) (M. Arjomand, Trans.). London/New York: Routledge.

Fiske, J. (1992). The cultural economy of fandom. In L. A. Lewis (Ed.), *The Adoring audience fan culture and popular media* (pp. 30–49). London/New York: Routledge.

Fong, T., Nourbakhsh, I., & Dautenhahn, K. (2003). A survey of socially interactive robots. *Robotics and Autonomous Systems, 42*(3–4), 143–166. doi:10.1016/S0921-8890(02)00372-X.

Ford, M. (2015). *Rise of the robots: Technology and the threat of a jobless future.* New York: Basic Books.

Freshwater, H. (2009). *Theatre & audience.* Houndmills/Basingstoke/Hampshire/New York: Palgrave Macmillan.

Fujita, N., & Kinase, A. (1991). The use of robotics in automated plant propagation. In I. K. Vasil (Ed.), *Scale-up and automation in plant propagation* (pp. 231–244). Cell Culture and Somatic Cell Genetics of Plants, Volume 8. San Diego/New York: Academic Press.

Funabiki, T. (2010). *Nihonjinron saikō* [A Reexamination of Nihonjinron]. Tokyo: Kōdansha.

Furuta, T. (2006). 2sai no korokara robotto hakase: Robotto gijutsu de bunmei, bunka no shinpo ni kōken sitai [Dr Robot from the age of two: Wanting to contribute to the progress of civilisation and culture through robot technology].

In P. H. P. Kenkyūjo (Ed.), *Otonano tameno robotto gaku* [Study of the robot for adults] (pp. 11–38). Tokyo: PHP Kenkyūjo.

Galbraith, P. W. (2011). Fujoshi: Fantasy play and transgressive intimacy among 'Rotten Girls' in contemporary Japan. *Signs, 37*(1), 211–232. doi:10.1086/660182.

Galbraith, P. W. (2012). Idols: The image of desire in Japanese consumer capitalism. In P. W. Galbraith & J. G. Karlin (Eds.), *Idols and celebrity in Japanese media culture* (pp. 185–208). Houndmills/Basingstoke/Hampshire/New York: Palgrave Macmillan.

Galbraith, P. W. (2015). *Otaku* sexuality in Japan. In M. J. McLelland & V. Mackie (Eds.), *Routledge handbook of sexuality studies in East Asia* (pp. 205–217). London/New York: Routledge.

Garner, S. B. (1994). *Bodied spaces: Phenomenology and performance in contemporary drama*. Ithaca: Cornell University Press.

Gell, A. (2009). The technology of enchantment and the enchantment of technology. In F. Candlin & R. Guins (Eds.), *The object reader* (pp. 208–228). London/New York: Routledge.

Geraci, R. M. (2010). *Apocalyptic AI visions of heaven in robotics, artificial intelligence, and virtual reality*. Oxford/New York: Oxford University Press.

Giannachi, G. (2004). *Virtual theatres: An introduction*. London/New York: Routledge.

Giannachi, G. (2007). *The politics of new media theatre: Life TM*. London/New York: Routledge.

Gibson, J. J. (1979). *The ecological approach to visual perception*. Boston: Houghton Mifflin.

Glasspool, L. (2012). From boys next door to boys' love: Gender performance in Japanese male idol media. In P. W. Galbraith & J. G. Karlin (Eds.), *Idols and celebrity in Japanese media culture* (pp. 113–130). Houndmills/Basingstoke/Hampshire/New York: Palgrave Macmillan.

Goldblatt, D. (2006). *Art and ventriloquism*. London/New York: Routledge.

Goodall, J. R. (2008). *Stage presence*. London/New York: Routledge.

Goodlander, J. (2015). Plaza Indonesia: Performing modernity in a shopping mall. In M. Omasta & D. Chappell (Eds.), *Play, performance, and identity: How institutions structure ludic spaces* (pp. 117–127). New York: Routledge.

Greenhalgh, P. (1988). *Ephemeral vistas: The expositions universelles, great exhibitions, and world's fairs, 1851–1939*. Manchester/New York: Manchester University Press.

Grehan, H. (2009). *Performance, ethics and spectatorship in a global age*. Basingstoke/New York: Palgrave Macmillan.

Grehan, H. (2015). Actors, spectators, and 'vibrant' objects: Kris Verdonck's ACTOR #1. *TDR: The Drama Review, 59*(3), 132–139. doi:10.1162/DRAM_a_00475.

Guillot, A., & Meyer, J. A. (2010). *How to catch a robot rat: When biology inspires innovation* (S. Emanuel, Trans.). Cambridge, MA: MIT Press.

Gundam Global Challenge. (2014, July 9). Talk about Gundam Global Challenge! A dream makes reality, and likewise reality makes a dream. *GGC Special Vol. 1.* http://gundam-challenge.com/en/. Accessed on 10-10-2015.

Gunkel, D. J. (2012). *The machine question critical perspectives on AI, robots, and ethics.* Cambridge, MA: MIT Press.

Haberman, C. (1985, May 5). Japanese see a "Made in Japan" future and feel reassured by that vision. *The New York Times*, sec. World. http://www.nytimes.com/1985/05/05/world/japanese-see-a-made-in-japan-future-and-feel-reassured-by-that-vision.html. Accessed on 10-12-2014.

Hakubutsukan, K. K. Shimbun, Y., & Nihon Terebi Hōsōmō (The National Science Museum, the Yomiuri Shimbun Publishing Company, and the Nippon Television Network Corporation). (2007). Goaisatsu [Greetings]. In K. Suzuki (Ed.), *Dai robotto haku: Karakuri kara anime, saishin robotto made* [The Great Robot Exhibition: From karakuri to anime to the latest robots] (p. 4), exh. cat. Tokyo: Yumiuri Shimbun.

Haldrup, M., & Larsen, J. (2010). *Tourism, performance and the everyday: Consuming the orient.* London/New York: Routledge.

Hamada, T., & Sano, T. (2011). Kōreisha serapī yō robotto no inshō ni kansuru chōsa [A study on impressions upon the elderly of therapy robots]. *Tsukubagakuin Daigakukiyō, 6*, 43–48.

Hamada, T., Hashimoto, T., Akazawa, T., & Matsumoto, Y. (2003). Robotto Serapī no Kanōsei ni Kansuru Ichi kōsatsu [A thought on possibilities for robot therapy]. *Kansei Testugaku, 3*, 98–109.

Hamano, S. (2008). *Ā kitekuchā no seitai kei: Jōhō kankyō wa ikani sekkei sarete kitaka* [Ecology of architecture: How the information environment has been designed]. Tokyo: NTT Shuppan.

Hanano Banpaku Kenbutsu Gaido Henshū Iinkai [The Editing Committee of the Guide Book for the Flower Expo]. (1990). *Expo '90 Hana no banpaku watashitachi no kenbutsu gaido* [Expo '90, the Flower Expo, Our Exhibition Guide]. Tokyo: Peppu Shuppan.

Harootunian, H. D. (2000a). *History's disquiet modernity, cultural practice, and the question of everyday life.* New York: Columbia University Press.

Harootunian, H. D. (2000b). *Overcome by modernity history, culture, and community in interwar Japan.* Princeton: Princeton University Press.

Harvie, J. (2009). *Theatre and the city.* Basingstoke/New York: Palgrave Macmillan.

Heim, C. (2016). *Audience as performer: The changing role of theatre audiences in the twenty-first century.* London/New York: Routledge.

Henderson, M. C. (2004). *The city and the theatre: The history of New York playhouses: A 250 year journey from Bowling Green to Times Square.* New York: Back Stage Books.

Hibino, K. (2012). Oscillating between Fakery and Authenticity: Hirata Oriza's Android Theatre. *Comparative Theatre Review, 11*(1), 30–42. doi:10.7141/ctr.11.30.

Hill, L., & Paris, H. (2006). *Performance and place*. Houndmills/Basingstoke/ Hampshire/New York: Palgrave Macmillan.

Hirakawa, S. (2006). *Wakon yōsai no keifu: Uchi to Soto kara no Meiji nihon, jō* [Genealogy of Japanese spirituality and Western Technology: Meiji Japan inside and outside, Part 1]. Tokyo: Heibonsha.

Hirata, O. (1995). *Gendai kōgo engeki no tameni* [For contemporary colloquial theatre]. Tokyo: Banseisha.

Hirata, O. (1998). *Engeki nyūmon* [Introduction to theatre]. Tokyo: Kōdansha.

Hirata, O. (2004). *Engi to enshutsu* [Acting and directing]. Tokyo: Kōdansha.

Hirata, O. (2012). About our Robot/Android Theatre (K. Hibino, Trans.). *Comparative Theatre Review, 11*(1): 29. doi:10.7141/ctr.11.29.

Hirata, O., Ishiguro, H., & Kinsui, S. (2010). Robotto ga engeki? Robotto to engeki!? [A robot in the theatre? Performing with a robot!?]. In Osaka daigaku komyunikēshon dezain sentā (Ed.), *Robotto Engeki*. (pp. 14–33). Osaka: Osaka Daigaku Shuppankai.

Hiromatsu, T. (1973). *Henkai no akusho* [Bad places at the margins of society]. Tokyo: Heibonsha.

Hirose, M. (2006). Nirin, yonrin, soshite kyūkyoku no idōtai e [Two wheels, four wheels, and the ultimate vehicle]. In P. H. P. Kenkyūjo (Ed.), *Otonano tameno robottogaku* [Study of the robot for adults] (pp. 113–141). Tokyo: PHP Kenkyūjo.

Hirose, S. (2011). *Robotto sōzōgaku nyūmon* [Introduction to a study on robot creation]. Tokyo: Iwanami shoten.

Hirose, M., & Ogawa, K. (2007). Honda humanoid robots development. *Philosophical Transactions of the Royal Society of London A: Mathematical, Physical and Engineering Sciences, 365*(1850), 11–19. doi:10.1098/rsta.2006.1917.

Hoggett, R. (2011, December 17). 1985: Marco and the Fuyo Robot Theater Expo'85 – Automax (Japanese). *Cyberneticzoo.com*. http://cyberneticzoo.com/ robots/1985-marco-and-the-fuyo-robot-theater-expo85-automax-japanese/. Accessed on 10-07-2014.

Honda, Y. (2014). *Robotto kakumei: naze gūguru to amazon ga tōshi suru noka* [The robot revolution: Why Google and Amazon have invested in it]. Tokyo: Shōdensha.

Honda Motor Company. (2015a). Honda ASIMO robotto kaihatsu no rekishi [The history of ASIMO's development]. http://www.honda.co.jp/ASIMO/history/. Accessed on 10-10-2015.

Honda Motor Company. (2015b). About Honda Robotics. http://world.honda.com/HondaRobotics/. Accessed on 10-10-2015.

Hornyak, T. N. (2006). *Loving the machine: The art and science of Japanese robots* (1st ed.). Tokyo/New York: Kodansha International.

Hotta, J. (2008). *Hito to robotto no himitu* [The secret concerning robots and humans]. Tokyo: Kōbunsha.

Huyssen, A. (1986). *After the great divide: Modernism, mass culture, postmodernism*. Bloomington: Indiana University Press.

Inami, M. (2015). 2020 nen wa chōjin orinpikkue no tsūkaten ni suginai [The Year 2020 is Merely an Episode in the Journey Toward a Superhuman Olympics]. Tsunehiro Uno (Ed.). *Planets, 9* (January): 58–63.

Inamura, T., & Ikeya, R. (2009). Nōkagaku to robotikusu [Brain science and robotics]. In T. Inamura, H. Sena, & R. Ikeya (Eds.), *Robotto no oheso* [The robot's navel] (pp. 48–65). Tokyo: Maruzen.

Inoue, H. (2007). *Nihon robotto sensōki* [History of Japan's Robot War] *1939–1945*. Tokyo: NTT Shuppan.

Ishida, M. (2008). Naka no hito ni naru: Koe modoki ga kanōni shita mono [Becoming an insider: What imitation voice can allow]. *Yuriika, 40*(15), 88–94.

Ishiguro, H. (2007). *Andoroido saiensu: Ningen wo sirutame no robotto kenkyū* [Android science: A study to learn what the human is]. Tokyo: Mainichi Komyunikēshonzu.

Ishiguro, H. (2009). *Robottoto wa nanika: Hito no kokoro wo utsusu kagami* [What is the robot?: A mirror reflecting the human soul]. Tokyo: Kōdansha Gendaishinsho.

Ishiguro, H. (2011). *Dōsureba hito wo tsukureruka: Andoroido ni natta watashi* [How can a human be made: I became an android]. Tokyo: Shinchōsha.

Ishiguro, H. (2012). *Hito to geijutsu to andoroido: Watashi wa naze robotto o tsukurunoka* [The human being, the arts, and the android: Why do I make androids?]. Tokyo: Nihonhyōronsha.

Ishiguro, H. (2013). *Kusobukuro no uchi to soto* [Inside and outside the human body]. Tokyo: Asahi Shimbun Shuppan.

Ishiguro, H., & Ikeya, R. (2010). *Robotto wa namida wo nagasuka* [Does a robot Shed Tears]. Tokyo: PHP Kenkyūjo.

Ishiguro, H., & Oriza, H. (2015). Ishiguro Hiroshi X Hirata Oriza: Aondoroido wa ningen no yume wo miruka [Can an android have dreams like humans do]. In Bungē Bessatsu (Ed.), *Hirata Oriza: Sōtokushū, sizukana kakumei no kishu* [Special issue, the leader of the quiet revolution] (pp. 118–125). Tokyo: Kawade shobō.

Ishiguro, H., & Washida, K. (2011). *Ikirutte nanyaroka: Kagakusha to tetsugakusha ga kataru, wakamono no tame no kuritikaru jinsei sinkingu* [What does it mean to live: A conversation between a scientist and philosopher, critical life thinking for young people]. Tokyo: Mainichi Shimbunsha.

Ishihara, N. (2011, April 28). Genpatus, anzen shinwa ni manshin sita tsumi [Nuclear power plants, A crime caused by a myth of safety]. *Nikkei Business Online.* http://business.nikkeibp.co.jp/article/manage/20110426/219655/. Accessed on 05-15-2012.

IT Media News. (2009, September 1). Jitsubutsudai Gandamu, kaitai stāto, raijōsha wa sanbai [Actual-sized Gundam, started to be dismantled, (results in) three times as many visitors]. http://www.itmedia.co.jp/news/articles/0909/01/news077.html. Accessed on 03-15-2012.

Ito, G. (2005). *Tezuka izu deddo: Hirakareta manga hyōgenron e* [Tezuka is dead: Open expression in manga]. Tokyo: NTT Shuppan.

Ito, M. (2012). Introduction. In M. Ito, D. Okabe, & I. Tsuji (Eds.), *Fandom unbound: Otaku culture in a connected world* (pp. xi–xxxi). New Haven: Yale University Press.

Iwabuchi, K. (2002). 'Soft' nationalism and narcissism: Japanese popular culture goes global. *Asian Studies Review, 26*(4), 447–469. doi:10.1080/10357820208713357.

Jackson, S. (2011). *Social works: Performing art, supporting publics.* New York: Routledge.

Japan Expo. (2014). The history of Japan Expo. http://www.japan-expo-paris.com/en/menu_info/history_475.htm. Accessed on 10-10-2015.

Japan National Tourism Organization. (2013). A city with two faces. *Discover the Sprit of Japan.* http://www.visitjapan.jp/en/m/player/?video=72. Accessed on 12-2-2013.

Jenkins, H. (2006). *Convergence culture: Where old and new media collide.* New York: New York University Press.

Jiji Tsūshinsha (Jiji News Service). (2015, December 18). 18 nen no nihon haku kentō: Abe shushō, bunka no miryoku hasshin, seifu kondankai [Considering the Japan Expo. 2018: Prime Minister Abe and the transmission of cultural power at a government round-table conference]. *Jiji Dotto Komu.* http://www.jiji.com/jc/zc?k=201512/2015121800946. Accessed on 01-15-2016.

Jiji Tsūshinsha (Jiji News Service). (2016, November 18). Gaikokujin kaigoshi wo zenmen kaikin: Kanren nihou seiritsu, jisshū mo ukeire [Lifting restrictions on foreign care workers: Two related laws passed; internships also permitted]. *Jiji Dotto Komu.* http://www.jiji.com/jc/article?k=2016111800056&g=pol. Accessed on 11-20-2016.

JIS Kikaku Yōgo (JIS Industrial Standard Terms). (2016). Robotto [Robot]. http://rbt.jisw.com. Accessed on 01-15-2016.

Kac, E. (2005). *Telepresence & bio art: Networking humans, rabbits & robots.* Ann Arbor: University of Michigan Press.

Kaerlein, T. (2015). Minimizing the human? Functional reductions of complexity in social robotics and their cybernetic heritage. In J. Vincent, S. Taipale, B. Sapio, G. Lugano, & L. Fortunati (Eds.), *Social robots from a human perspective* (pp. 77–88). Cham: Springer International Publishing.

Kajita, S. (2008). Hyūmanoido robotto kenyū no genba yori [On the ground with humanoid research]. In S. Hideaki (Ed.), *Saiensu imajinēshon: Kagaku to SF no saizensen, soshite miraie* [Scientific imagination: The frontline of science and SF, and the future] (pp. 38–53). Tokyo: NTT Shuppansha.

Kakoudaki, D. (2014). *Anatomy of a robot: Literature, cinema, and the cultural work of artificial people.* New Brunswick/New Jersey/London: Rutgers University Press.

Kameda, T. (2015). Komyunikēshon robotto wo mochiita chiiki fukushi shisetsu deno rekurēshon jisshi no bunseki [A study of the use of communication robots in recreation activities at regional care facilities]. *Sōkajoshi Tankidaigaku Kiyō 46* (February), 9-23.

Kanamori, M., Suzuki, M., Oshiro, H., Tanaka, M., Inoguchi, T., Takasugi, H., Saito, Y., & Yokoyama, T. (2003). Pilot study on improvement of quality of life among elderly using a pet-type robot. In *Proceedings of IEEE international symposium on computational intelligence in robotics and automation* (Vol. 1, 107–112). doi:10.1109/CIRA.2003.1222072.

Kanda, T., & Ishiguro, H. (2006). An approach for a social robot to understand human relationships: Friendship estimation through interaction with robots. *Interaction Studies, 7*(3), 369–403. doi:10.1075/is.7.3.12kan.

Kang, M. (2011). *Sublime dreams of living machines the automaton in the European imagination.* Cambridge, MA/London: Harvard University Press.

Kaplan, F. (2011). *Robotto wa tomodachi ni nareruka: nihonnjin to kikai no fusigi na kankei* [Can a robot be our friend: The mysterious relationship between the Japanese and robots]. Translated from French to Japanese by Kenji Nishi. [*Les Machines Apprivoisées: Comprendre les Robots de Loisir*]. Tokyo: NTT Shuppan.

Kato, N. (2012). Robotto wo rihabiritēshon ni dōnyū shitemite omou koto [A thought on the introduction of robots to rehabilitation]. *Keisokuto Seigyo, 51*(7), 605–608.

Katsuki, T. (2014). *Aidoru no yomikata* [A way to read idols]. Tokyo: Seikyūsha.

Kaye, N. (2000). *Site-specific art performance, place, and documentation.* London/New York: Routledge.

Kennedy, D. (2009). *The spectator and the spectacle: Audiences in modernity and postmodernity.* Cambridge/New York: Cambridge University Press.

Kijima, Y. (2012). The fighting gamer otaku community: What are they 'fighting' about?. In M. Ito, D. Okabe, & I. Tsuji (Eds.), *Fandom unbound: Otaku culture in a connected world* (pp. 249–274). New Haven: Yale University Press.

Kimura, R. (2012). Pettogata robotto wo mochi'ite ninchi kōreisha wo taishōnisita robotto serapī [Therapy for the cognitively-impaired elderly with pet-like robots]. *Keisokuto Seigyo, 51*(7), 605–608.

Kinsella, S. (2000). *Adult manga: Culture and power in contemporary Japanese society.* Richmond/Surrey: Curzon.

Kishi, N. (2011). *Robotto ga nihon o suku'u* [Robots will save Japan]. Tokyo: Bungē shunjū.

Klich, R., & Scheer, E. (2012). *Multimedia performance.* Houndmills/Basingstoke/Hampshire/New York: Palgrave Macmillan.

Klingmann, A. (2007). *Brandscapes: Architecture in the experience economy.* Cambridge, MA: MIT Press.

Kobayashi, H. (2006). *Robotto sinkaron: Jinzōningen kara hito to kyōzon suru sisutemu e* [Robot evolution: From Jinzōningen to a system for coexistence with humans]. Tokyo: Ōmusha.

Konijn, E. (2000). *Acting emotions: Shaping emotions on stage* (B. Leach, Trans.). Amsterdam: Amsterdam University Press.

Kubo, T. (2011). Special interview. *Inter-X-Cross Creative Center.* http://www. icc-jp.com/special/2011/01/001726.php. Accessed on 04-15-2013.

Kubo, A. (2015). *Robotto no jinruigaku: Nijūseiki no kikai to ningen* [Anthropology of the robot: The machine and the human in the 20th century]. Tokyo: Sekai Sisōsha.

Kuniyoshi, Y. (2008). Robotto bodei robotto maindo: Mono no sekai to kokoro no sekai no yūgō soshite kaihatsu [Robot body, robot mind: Developing the integration of the worlds of soul and object]. In H. Sena (Ed.), *Saiensu imajinēshon: kagaku to SF no saizensen, soshite miraie* [Scientific imagination: The frontline of science and SF, and the future] (pp. 70–87). Tokyo: NTT Shuppansha.

Kurata, K. (2012). KURATAS. *Suidobashi Jūkō.* http://suidobashijuko.jp/index. php. Accessed on 03-10-2014.

Kurata, K. (2015a). Response to robot duel challenge. *Suidobashi Jūkō.* http:// suidobashijuko.jp/index.php. Accessed on 12-10-2015.

Kurata, K. (2015b, July 23). Kuratasu vs Megabotto, sono 2 [Kuratas vs Megabot, Part 2]. Nandemo tsukuruyo: Hontoni ugokuka kyodai robo [I would make anything: Does the giant robot really move?] (blog). http://monkeyfarm. cocolog-nifty. com/nandemo/2015/07/post-5853.html. Accessed on 10-08-2015.

Kurata, K. (2016, September 6). Kinkyō houkoku dado [Latest news]. Nandemo tsukuruyo: Hontoni ugokuka kyodai robo [I would make anything: Does the giant robot really move?] (blog). http://monkeyfarm.cocolog-nifty.com/ nandemo/2016/09/post-31bf.html. Accessed on 10-20-2016.

Kuroki, K. (2010). Robotto engeki no kaihatsu [The development of robot theatre]. In Osaka daigaku komyunikēshon dezain sentā (Ed.), *Robotto engeki* [Robot theatre] (pp. 60–65). Osaka: Osaka Daigaku Shuppankai.

LaMarre, T. (2009). *The anime machine: A media theory of animation.* Minneapolis: University of Minnesota Press.

Lash, S., & Lury, C. (2007). *Global culture industry: The mediation of things.* Cambridge: Polity.

Latour, B. (1996). On interobjectivity (G. Bowker, Trans.). *Mind, Culture, and Activity, 3*(4), 228–245. doi:10.1207/s15327884mca0304_2.

Levinas, E. (1991). *Otherwise than being, or, beyond essence* (A. Lingis, Trans.). Dordrecht/Boston/London: Kluwer Academic Publishers.

Levy, D. N. L. (2007). *Love and sex with robots: The evolution of human-robot relations.* New York: HarperCollins.

Lin, P., Abney, K., & Bekey, G. A. (2012). *Robot ethics: The ethical and social implications of robotics.* Cambridge, MA: The MIT Press.

Loxley, J. (2007). *Performativity.* London/New York: Routledge.

MacCannell, D. (1973). Staged authenticity: Arrangements of social space in tourist settings. *American Journal of Sociology, 79*(3), 589–603.

MacDorman, K. F., Vasudevan, S. K., & Ho, C.-C. (2009). Does Japan really have robot mania? Comparing attitudes by implicit and explicit measures. *AI & Society, 23*(4), 485–510. doi:10.1007/s00146-008-0181-2.

Mackintosh, I. (1993). *Architecture, actor, and audience*. London/New York: Routledge.

Masuda, S. (2008). Hatsune miku kara tōku hanarete [Away from Hatsune Miku]. *Yuriika, 40*(15), 184–192.

Matsubara, H. (1999). *Tetsuwan Atomu wa jitsugen dekiruka: Robo kappu ga hiraku mirai* [Would we be able to actualise astro boy?: The future created by RoboCup]. Tokyo: Kwade Shobō.

Matsubara, H. (2001). Robo kappu no yume [A dream of RoboCup]. In H. Matsubara, I. Takeuchi & H. Numata (Eds.), *Robotto no jōhōgaku: 2050 nen wārudo kappu ni katsu* [Robot information study: Winning the World Cup in 2050] (pp. 7–66). Tokyo: NTT Shuppansha.

Matsui, Y. (2012, October 5). Futuristic bot cabaret wows Tokyo. *The Japan Times Online*. http://www.japantimes.co.jp/news/2012/10/05/national/futuristic-bot-cabaret-wows-tokyo/. Accessed on 01-15-2013.

Matsumoto, K. [Katsuya] (2015). *Hirata Oriza: Sizukana engeki to iu hōhō* [The methodology of quiet theatre]. Tokyo: Sairyūsha.

Matsumoto, M. (2015, June 1). Petto robo AIBO no shi wo nageki kanashimu kōreisha tachi gijutsusha OB ga oishasan ni nanori wo Ageru [The elderly who suffer grief over the death of their pet robot AIBO, an OB (retired man) technician stands in as a doctor]. *Kyarikone Nyūsu*. https://news.careerconnection.jp/?p=12309. Accessed on 11-20-2015.

Mattie, E. (1998). *World's fairs*. New York: Princeton Architectural Press.

Maxwell, I. (2008). The ritualization of performance (studies). In G. St. John (Ed.), *Victor Turner and contemporary cultural performance* (pp. 59–75). New York/Oxford: Berghahn Books.

McAuley, G. (1999). *Space in performance: Making meaning in the theatre*. Ann Arbor: University of Michigan Press.

McAuley, G. (Ed.). (2006). *Unstable ground: Performance and the politics of place*. Bruxelles/Oxford: P.I.E. Peter Lang.

McConachie, B. A. (2008). *Engaging audiences: A cognitive approach to spectating in the theatre*. New York/Basingstoke: Palgrave Macmillan.

McKinnie, M. (2007). *City stages: Theatre and urban space in a global city*. Toronto: University of Toronto Press.

McLelland, M., Suganuma, K., & Welker, J. (2007). Introduction: Re(claiming) Japan's queer past. In M. McLelland, K. Suganuma, & J. Welker (Eds.), *Queer voices from Japan: First person narratives from Japan's sexual minorities* (pp. 1–29). Lanham: Lexington Books.

McVeigh, B. (1996). Commodifying affection, authority and gender in the everyday objects of Japan. *Journal of Material Culture, 1*(3), 291–312. doi:10.1177/135918359600100302.

Meadows, M. S. (2011). *We, robot: Skywalker's hand, blade runners, Iron Man, slutbots, and how fiction became fact*. Guilford: Lyons Press.

Menzel, P., & D'Aluisio, F. (2000). *Robo sapiens: Evolution of a new species.* Cambridge, MA: MIT Press.

Miles, M. (1997). *Art, space and the city: Public art and urban futures.* London/ New York: Routledge.

Miller, L. (2004). You are doing Burikko!: Censoring/scrutinizing artificers of cute femininity in Japanese. In S. Okamoto & J. S. Shibamoto Smith (Eds.), *Japanese language, gender, and ideology cultural models and real people* (pp. 148–165). New York: Oxford University Press.

Miller Frank, F. (1995). *The mechanical song: Women, voice, and the artificial in nineteenth-century French narrative.* Stanford: Stanford University Press.

Mishima, T. (Ed.). (2005). *Toyota Group Pavilion Official Guide Book.* Toyota Group.

Miyoshi, M., & Harootunian, H. D. (1989). Introduction. In M. Miyoshi & H. D. Harootunian (Eds.), *Postmodernism and Japan* (pp. vii–xix). Durham: Duke University Press.

Mizu'ushi, K. (2011, October 5). Hirata Oriza X Ishiguro kenkyūshitsu, Andoroido engeki, Sayonara [Hirata Oriza X Ishiguro Laboratory, Andoroido Theatre, Sayonara]. Wonderland. http://www.wonderlands.jp/archives/18906/.

Mochizuki, T., & Pfanner, E. (2015, February 11). In Japan, dog owners feel abandoned as Sony stops supporting "Aibo." *Wall Street Journal.* http://www. wsj.com/articles/in-japan-dog-owners-feel-abandoned-as-sony-stops-supporting-aibo-1423609536. Accessed on 11-15-2015.

Mori, M. [Masahiro]. (1981). *The Buddha in the robot* (C. S. Terry, Trans.). Tokyo: Kōsei Publishing.

Mori, M. [Masahiro]. (1999). *Robo kon hakase no monotsukuri yūron* [Dr Robocon on the art of making]. Tokyo: Ōmusha.

Mori, M. [Masahiro]. (2012). The uncanny valley [From the field] (K. F. MacDorman & N. Kageki, Trans.). *IEEE Robotics Automation Magazine, 19*(2): 98–100. doi:10.1109/MRA.2012.2192811.

Mori, M. [Masahiro]. (2014). *Robotto kōgaku to ningen: Mirai no tameno robotto kōgaku* [Robotics and the human: Robotics for the future]. Tokyo: Ōmusha.

Mori, M. (2002). The structure of theater: A Japanese view of theatricality. *Substance, 31*(2), 73–93. doi:10.1353/sub.2002.0033.

Mori, S., Sano, M., Adachi, N., & Gao, J. (2014). Ningen to robotto no jōdō intarakushon ni okeru tekiō teki hikikomi seigyo to gakushū [Adaptive control and learning of entrainment in affective interaction between humans and robots]. *Intarakushon 2014 Ronbunshū,* February, 232–233.

Morikawa, K. (2008). *Shuto no tanjō: Moeru toshi Akihabara, zōhoban* [The birth of Hobby city: 'Moe' city Akihabara, an expanded edition]. Tokyo: Gentōsha.

Morikawa, K. (2011). Azuma Hideo wa ikanishite otaku bunka no so ni nattaka [How did Hideo Azuma become the founder of *otaku* culture]. In *Azuma Hideo: Bishōjo, SF, fujōri gyagu, soshite shissō, sōtokushū* [Beautiful girl, SF, nonsense gag, and disappearance: A special issue] (pp. 179–186). Kawade Yumemukku Bungei Bessatsu. Tokyo: Kawade Shobō Shinsha.

Moriyama, K. (2007, October 23). Kokuritsu kagaku hakubutsukan, dai robottohaku wo kaisai: ASIMO kara saishin robotto kara, karakuri, anime made [The opening of the great robot exhibition at the national museum of nature and science: From ASIMO to the latest robots, (from) *karakuri* to anime]. *Robot Watch.* http://robot.watch.impress.co.jp/cda/news/2007/10/23/704.html. Accessed on 01-15-2008.

Moriyama, K. (2009, June 17). Nihon Jidō bunka kenyūjo, Aizawa Jirō seisaku no robotto no fukugen sagyō wo kaishi [The Japan Institute of Juvenile Culture, Commencement of the Reconstruction of Jiro Aizawa's Robots]. *Robot Watch.* http://robot.watch.impress.co.jp/docs/news/20090617_294254.html. Accessed on 11-30-2009.

Morley, D., & Robins, K. (1995). *Space of identity: Global media, electronic landscapes and cultural boundaries.* London/New York: Routledge.

Morse, M. (2016). *Soft is fast: Simone Forti in the 1960s and after.* Cambridge, MA: MIT Press.

Moyle, W., Beattie, E., Draper, B., Shum, D., Thalib, L., Jones, C., O'Dwyer, S., & Mervin, C. (2015). Effect of an interactive therapeutic robotic animal on engagement, mood states, agitation and psychotropic drug use in people with dementia: A cluster-randomised controlled trial protocol. *BMJ Open, 5*(8), e009097. doi:10.1136/bmjopen-2015-009097.

Munroe, A. (2005). Introducing little boy. In T. Murakmi (Ed.), *Little boy: The arts of Japan's exploding subculture* (pp. 241–261). New York/New Haven/London: Japan Society and Yale University Press.

Murakami, T. (2005). Earth in my window. In T. Murakami (Ed.), *Little boy: The arts of Japan's exploding subculture* (pp. 99–149). New York/New Haven/London: Japan Society and Yale University Press.

Murata, J. (2003). Creativity of technology: An origin of modernity?. In T. J. Misa, P. Brey, & A. Feenberg (Eds.), *Modernity and technology* (pp. 151–177). Cambridge, MA: MIT Press.

Nagaike, K. (2012). Johnny's idols as icons: Female desires to fantasize and consume male idol images. In P. W. Galbraith & J. G. Karlin (Eds.), *Idols and celebrity in Japanese media culture* (pp. 97–112). Houndmills/Basingstoke/Hampshire/New York: Palgrave Macmillan.

Nagata, T. (2005). *Robotto wa ningen ni nareruka* [Can a robot be a human being]. Tokyo: PHP Kenkūsho.

Najita, T. (1989). On culture and technology in postmodernism and Japan. In M. Miyoshi & H. D. Harootunian (Eds.), *Postmodernism and Japan* (pp. 3–20). Durham: Duke University Press.

Nakamura, M. (2007). Horror and machines in Prewar Japan: The mechanical Uncanny in Yumeno Kyūsaku's Dogura Magura. In C. Bolton, I. Csicsery-Ronay, & T. Tatsumi (Eds.), *Robot ghosts and wired dreams Japanese science fiction from origins to anime* (C. Bolton, Trans.) (pp. 3–26). Minneapolis: University of Minnesota Press.

Nakanishi, O. (2014). Kurosu rebyū: Hirata Oriza + Ishiguro kenyūshitsu, andoroido engeki, Sayonara [Cross-review: Hirata Oriza + Ishiguro Laboratory, Android Theatre, Sayonara]. *Kokusai Engekika Kyōkai Engeki Hyōronshi, ACT* 21. http://act-kansai.net/?p=73. Accessed on 10-02-2015.

Nakayama, S. (2006). *Robotto ga nihon o suku'u* [Robots will save Japan]. Tokyo: Toyō Keizai Shinpōsha.

Napier, S. J. (1993). Panic sites: The Japanese imagination of disaster from Godzilla to Akira. *Journal of Japanese Studies, 19*(2), 327–351. doi:10.2307/132643.

National Diet Library. (2010). First National Industrial Exhibition. http://www. ndl.go.jp/exposition/e/s1/naikoku1.html. Accessed on 03-03-2015.

National Institute of Advanced Industrial Science and Technology. (2003). Hataraku ningengata robotto: Ningen kyōchō, kyōzongata robotto sisutemu no kenkyū kaihatsu wo happyō [Announcement on the research concerning the robot system that cooperates and cohabits with humans]. http://www.aist.go.jp/aist_j/press_release/pr2003/pr20030226/pr20030226.html. Accessed on 05-07-2014.

National Institute of Informatics. (2011). Creating robot theater for building a more preferable robot. *KAKEN: Database of Grants-in-Aid for Scientific Research.* https://kaken.nii.ac.jp/d/p/23240027.ja.html. Accessed on 07-15-2014.

Natsume, F. (1995). *Tezuka Osamu wa dokoni iru* [Where is Osamu Tezuka]. Tokyo: Chikuma Shobō.

NEDO Books henshū iinkai (NEDO Book Editing Committee). (2009). *RT supirittsu: Hito ni yakudatsu robotto gijutsu o kaihatsu suru* [RT spirits: Developing useful robots for humans]. Kawasaki: New Energy and Industrial Technology Development Organization.

Nelson, V. (2001). *The secret life of puppets.* Cambridge, MA: Harvard University Press.

Nestruck, J. K. (2013, February 27). Sayonara, I, worker: These plays are a little too robotic. *The Globe and Mail.* http://www.theglobeandmail.com/arts/theatre-and-performance/theatre-reviews/sayonara-i-worker-these-plays-are-a-little-too-robotic/article9123716/. Accessed on 12-10-2014.

News Limited. (2013). Japan AKB48 Pop Idol Minami Minegishi Shaves Head in Penance for Spending Night with Man. *news.com.au.* http://www.news.com.au/entertainment/music/japan-akb48-pop-idol-minami-minegishi-shaves-head-in-penance-for-spending-night-with-man/story-e6frfn09-1226567125292. Accessed on 10-04-13.

NHK Sābisu Sentā [NHK Service Centre]. (1985). *Nijū isseki ga mieta kagaku banpaku 85* [The 21st century is in sight at expo '85] (VHS). Tokyo: Tōhō.

Nihon Kokusai Hakurankai Kyōkai [The Japan International Exposition Committee]. (2005). *2005 nen nihon kokusai hakurankai kyōkai: Ai chikyūhaku kōshiki gaido bukku* [2005 Japan international exposition committee: Love the earth expo official guide book]. Tokyo: Pia.

Nimiya, K. (2009). Takarazuka to shōjo to moe [Takarazuka, Girls, and 'Moe']. In Sekyūsha Henshūbu (Ed.), *Takarazuka toiu sōchi* [Takarazuka as an apparatus] (pp. 307–340). Tokyo: Sekyūsha.

Nina, T. (2013, December 12). All the crazy stereotypes of Japan in one show. Comments on TripAdvisor. *Robot Restaurant*. http://www.tripadvisor.com. au/Attraction_Review-g1066457-d4776370-Reviews-or20-Robot_Restaurant-Shinjuku_Tokyo_Tokyo_Prefecture_Kanto.html#REVIEWS. Accessed on 12-20-2013.

Ninomiya, T. (2015). Komyunikēshon robotto Palro no shōkai to sagami robotto sangyōku ni okeru torikomi [Introducing communication robot Palro and activities at a robot operation area in Sagami]. *Nihon Robotto Gakkaishi, 33*(8), 607–610. doi:10.7210/jrsj.30.1000.

Nishida, T. (2005). Riaru ējento to siteno robotto [A robot as a real agent]. In H. Inoue, T. Kanede, M. Uchiyama, M. Asada, & Y. Anzai (Eds.), *Iwanami kōza robottogaku 5: Robotto infomatikkusu* [Iwanami robot science 5: Robot informatics] (pp. 89–139). Tokyo: Iwanami Shoten.

Nishida, K. (2007). *Essensharu Nishida, soku no kan, Nishida Kitaro kīwādo ronshū* [Essential Nishida: An issue on instantaneity: Essays that contain Kitaro Nishida's Keywords]. Tokyo: Shoshi Shinsui.

Norman, D. A. (1988). *The psychology of everyday things*. New York: Basic Books.

Norman, D. A. (2004). *Emotional design: Why we love (or hate) everyday things*. New York: Basic Books.

Nukada, H., & Murakami, T. (2005). Little boy (plates and entries). In T. Murakami (Ed.), *Little boy: The arts of Japan's exploding subculture* (pp. 1–97). New York/New Haven/London: Japan Society and Yale University Press.

Numata, H. (2001). Robokappu no sozōryoku [RoboCup imagination]. In H. Matsubara, I. Takeuchi, & H. Numata (Eds.), *Robotto no jōhōgaku: 2050 nen wārudo kappu ni katsu* [Robot information study: Winning the World Cup in 2050] (pp. 117–155). Tokyo: NTT Shuppansha.

Oba, M. (2013a). Cinema in the wood. In K. Yanobe (Ed.), *Yanobe Kenji: 1969–2005* (p. 111). Kyoto: Seigensha.

Oba, M. (2013b). The ruins of the future. In K. Yanobe (Ed.), *Yanobe Kenji: 1969–2005* (p. 67). Kyoto: Seigensha.

Oddey, A., & White, C. (2009). *Modes of spectating*. Bristol/Chicago: Intellect.

OED (Online). (2000). Oxford: Oxford University Press.

Ogasawara, T., Tajima, T., Hatakeyama, M., & Nishida, T. (2004). Hikikomi genshō ni motozuku ningen to robotto no anmoku jōhō no komyunikēshon [Communication of unconscious information between robots and humans based on the phenomenon of entrainment]. *Jinkōchinō gakkai Zenkoku taikai Ronbunshū* JSAI04: 115–115. doi:10.11517/pjsai.JSAI04.0.115.0.

Ohnuki-Tierney, E. (1987). *The monkey as mirror: Symbolic transformations in Japanese history and ritual*. Princeton: Princeton University Press.

Okada, M. (2012). *Yowai robotto* [Weak robot]. Tokyo: Igaku Shoin.

Okada, M., & Matsumoto, K. [Kotaro] (2014). Prorōgu [Prologue]. In M. Okada & K. [Kotaro] Matsumoto (Eds.), *Robotto no kanashimi: Komyunikēshon wo meguru hito to robotto no seitaigaku* [Sorrow of the robot: Ecology of communication between humans and robots] (pp. i–v). Tokyo: Shinyōsha.

Okada, T., Morikawa, K., & Murakami, T. (2005). Otaku talk. In T. Murakami (Ed.), *Little boy: The arts of Japan's exploding subculture* (Reiko Tomii, Trans.) (pp. 165–185). New York/New Haven/London: Japan Society and Yale University Press.

Okuno, T. (2002). *Ningen dōbutsu kikai: Tekuno animizumu* [Human, animal, machine: Techno animism]. Tokyo: Kadokawa shoten.

Omori, S. (1981). *Nagare to yodomi: Tetsugaku danshō* [Flow and stagnation: Philosophical fragments]. Tokyo: Sangyō Tosho.

Orbaugh, S. (2005). The genealogy of the cyborg in Japanese popular culture. In K.-y. Wong, G. Westfahl, & A. K.-s. Chan (Eds.), *World weavers globalization, science fiction, and the cybernetic revolution* (pp. 55–71). Hong Kong: Hong Kong University Press.

Orbaugh, S. (2007). Sex and the single cyborg: Japanese popular culture experiments in subjectivity. In C. Bolton, I. Csicsery-Ronay, & T. Tatsumi (Eds.), *Robot ghosts and wired dreams Japanese science fiction from origins to anime* (Christopher Bolton, Trans.). (pp. 222–249). Minneapolis: University of Minnesota Press.

Orikuchi, S. (1991). *Nihon geinōshi rokkō* [The history of Japanese entertainment: Six lectures]. Tokyo: Kōdansha.

Osakafu Nihon Banpaku Kinen Kōen Jimusho (Expo '70 Commemorative Park Office). (2014). Taiyō no tō [The Tower of the Sun]. *Banpaku Kinen Kōen.* http://www.expo70-park.jp. Accessed on 15-10-2014.

Osawa, M. (2008). *Fukanōsei no jidai* [The age of impossibility]. Tokyo: Iwanami Shoten.

Otsuka, E. (1997). *Shōjo minzokugaku: Seikimatsu no shinwa wo tsumugu, miko no matsuei* [Ethnology of the girl: Spinning a myth at the end of the century, descendants of the Japanese Shrine Maiden]. Tokyo: Kōbunsha.

Otsuka, E. (2001). Azuma Hideo: Otakunaru mono no kigen [Hideo Azuma: The origin of the *otaku*]. In E. Otsuka & G. Sasakibara (Eds.), *Kyōyō toshite no manga, anime* [Manga, anime as education] (pp. 85–108). Tokyo: Kōdansha Gendai Shinsho.

Otsuka, E. (2009). *Atomu no meidai: Tezuka Osamu to sengo manga no shudai* [An atom thesis: Osamu Tezuka and themes in Postwar Manga]. Tokyo: Kadokawa shoten.

Otsuka, E. (2012). *Monogatari shōhiron kai* [A theory of narrative consumption, revised edition]. Tokyo: Kadokawa Gurūpu Paburisshingu.

Otsuka, M. (2016, February 24). 2020 nen no robotto orinpikku wa dōnaru?: Hatsuno shimonkaigi ga kaisai [What could happen to the 2020 Olympics: The first advisory committee meeting was held]. *Mainabi Nyūsu.* http://news.mynavi.jp/articles/2016/02/04/robot_olympic/. Accessed on 03-03-2016.

Otsuka, E., & Osawa, N. [Nobuaki] (2005). *Japanimeishon wa naze yabureruka* [Why 'Japanimation' will be defeated]. Tokyo: Kadokawa Shoten.

Parker-Starbuck, J. (2011). *Cyborg theatre: Corporeal/technological intersections in multimedia performance*. Houndmills/Basingstoke/Hampshire/New York: Palgrave Macmillan.

Parker-Starbuck, J. (2015). Cyborg. Returns: Always-already subject technologies. In S. Bay-Cheng, J. Parker-Starbuck, & D. Z. Saltz (Eds.), *Performance and media: Taxonomies for a changing field* (pp. 65–92). Ann Arbor: University of Michigan Press.

Pearson, M. (2010). *Site-specific performance*. Houndmills/Basingstoke/Hampshire/New York: Palgrave Macmillan.

Pearson, M., & Shanks, M. (2001). *Theatre/archaeology*. London/New York: Routledge.

Perkowitz, S. (2004). *Digital people: From bionic humans to androids*. Washington, DC: Joseph Henry Press.

Posner, D. N. (2014). Contemporary investigations and hybridizations. In D. N. Posner, C. Orenstein, & J. Bell (Eds.), *The Routledge companion to puppetry and material performance* (pp. 225–227). London/New York: Routledge.

Poulton, C. (2014). From puppet to robot: Technology and the human in Japanese theatre. In D. N. Posner, C. Orenstein, & J. Bell (Eds.), *The Routledge companion to puppetry and material performance* (pp. 280–293). London/New York: Routledge.

Power, C. (2008). *Presence in play a critique of theories of presence in the theatre*. Amsterdam/New York: Rodopi.

Pyne, L. (2014, September 30). The day we brought our robot home. *The Atlantic*. http://www.theatlantic.com/technology/archive/2014/09/the-day-we-brought-our-robot-home/380891/. Accessed on 10-03-2015.

Rabbitt, S. M., Kazdin, A. E., & Scassellati, B. (2015). Integrating socially assistive robotics into mental healthcare interventions: Applications and recommendations for expanded use. *Clinical Psychology Review, 35*, 35–46. doi:10.1016/j.cpr.2014.07.001.

Radbourne, J., Glow, H., Johanson, K., Thomas, E., & Marshall, M. (Eds.). (2013). *The audience experience: A critical analysis of audiences in the performing arts*. Bristol/Chicago: Intellect.

Rancière, J. (2009). *The emancipated spectator* (G. Elliott, Trans.). London: Verso.

Raz, J. (1983). *Audience and actors: A study of their interaction in the Japanese traditional theatre*. Leiden: E.J. Brill.

Reilly, K. (2011). *Automata and mimesis on the stage of theatre history*. Houndmills/Basingstoke/Hampshire/New York: Palgrave Macmillan.

Reinelt, J. G. (2002). The politics of discourse: Performativity meets theatricality. *SubStance, 31*(2), 201–215. doi:10.1353/sub.2002.0037.

Rickly-Boyd, J. M. (2013). Existential authenticity: Place matters. *Tourism Geographies, 15*(4), 680–686. doi:10.1080/14616688.2012.762691.

Riek, L. D. (2012). Wizard of Oz studies in HRI: A systematic review and new reporting guidelines. *Journal of Human-Robot Interaction, 1*(1), 119–136. doi:10.5898/JHRI.1.1.Riek.

Roach, J. R. (2007). *It*. Ann Arbor: University of Michigan Press.

Robertson, J. (2007). Robo sapiens Japanicus: Humanoid robots and the posthuman family. *Critical Asian Studies, 39*(3), 369–398. doi:10.1080/14672710701527378.

Robertson, J. (2010). Gendering humanoid robots: Robo-sexism in Japan. *Body & Society, 16*(2), 1–36. doi:10.1177/1357034X10364767.

Robertson, J. (2014). Human rights vs. robot rights: Forecasts from Japan. *Critical Asian Studies, 46*(4), 571–598. doi:10.1080/14672715.2014.960707.

Roche, M. (2000). *Mega-events and modernity: Olympics and expos in the growth of global culture*. London: Routledge.

Rocket News 24. (2012, July 30). Mirai kita! Hito ga notte sōjū dekiru kyodai robotto, Kuratasu ga tsuini ohirome! Nanto hanbai mo kaishi [The future is here! The showing of the giant rideable robot Kuratas! Its sale has begun]. http://rocketnews24.com/2012/07/30/235808/. Accessed on 12-15-2013.

Rodogno, R. (2015). Social robots, fiction, and sentimentality. *Ethics and Information Technology*, 1–12. doi:10.1007/s10676-015-9371-z.

Rydell, R. W. (1984). *All the world's a fair: Visions of empire at American international expositions, 1876–1916*. Chicago: University of Chicago Press.

Šabanović, S. (2014). Inventing Japan's 'robotics culture': The repeated assembly of science, technology, and culture in social robotics. *Social Studies of Science, 44*(3), 342–367. doi:10.1177/0306312713509704.

Saijo, T. (2012, July). Hito ga notte sōjū dekiru kyodai robotto, Kuratasu [Kuratas, A giant robot that can be piloted]. *Wired Japan*. http://wired.jp/2012/07/26/kuratas/. Accessed on 12-15-2013.

Saito, T. (2007). Otaku sexuality. In C. Bolton, I. Csicsery-Ronay, & T. Tatsumi (Eds.), *Robot ghosts and wired dreams Japanese science fiction from origins to anime* (C. Bolton, Trans.) (pp. 222–249). Minneapolis: University of Minnesota Press.

Saito, T. (2011a). *Beautiful fighting girl* (J. K. Vincent & D. Lawson, Trans.). Minneapolis: The University of Minnesota Press.

Saito, T. (2011b). *Kyrakutā seishin bunseki: manga, bungaku, nihonjin* [*Character psychoanalysis: Manga, literature, the Japanese*]. Tokyo: Chikuma Shobō.

Sakakibara Kikai. (2016). Landwalker. http://www.sakakibara-kikai.co.jp/products/other/LW.htm. Accessed on 30-01-2016.

Sakurai, Y. (2007). *Firosofia robotika: Ningenni chikazuku robotto ni chikazuku ningen* [Philosophia robotica: Humans who become similar to robots that approximate humans]. Tokyo: Mainichi Komyunikēshonzu.

Salter, C. (2010). *Entangled: Technology and the transformation of performance*. Cambridge, MA: MIT Press.

Saltz, D. Z. (2015). Sharing the stage with media: A taxonomy of performer-media interactions. In S. Bay-Cheng, J. Parker-Starbuck, & D. Z. Saltz (Eds.), *Performance and media: Taxonomies for a changing field* (pp. 93–125). Ann Arbor: University of Michigan Press.

Sandry, E. (2015). *Robots and communication*. New York: Palgrave Macmillan.

Sankai, Y. (2006). Shintaikinō wo kakuchō suru: Robotto sūtsu de hito no yakunitatsu [Expanding physical capabilities: A robot-suit that is useful for humans]. In P. H. P. Kenkyūjo (Ed.), *Otonano tameno robotto gaku* [Study of the robot for adults] (pp. 93–114). Tokyo: PHP Kenkyūjo.

Sasaki, A. (2011). *Sokkyō no kaitai/kaitai: Ensō to engeki no aporia* [Dismantling and conceiving improvisation: Aporia in playing musical instruments and in theatre]. Tokyo: Seidosha.

Sasakibara, G. (2004). *Bishōjo no gendaishi: Moe to kyarakutā* [A modern history of the beautiful girl: 'Moe' and character]. Tokyo: Kōdansha.

Sawaragi, N. (2002a). *Expose 2002: Far beyond Dreams of the Future, Kenji Yanobe X Arata Isozaki.*

Sawaragi, N. (2002b). Isozaki Arata Intabyu: 1970 nen, Osaka banpaku wo kataru [Arata Isozaki interview: A talk on the 1970 Osaka Expo]. In Kirin Puraza Osaka (Ed.), *Expose 2002* (pp. 10–11). Osaka: Kirinbīru KPO Kirin Puraza Osaka.

Sawaragi, N. (2005a). On the battlefield of 'Superflat': Subculture and art in postwar Japan. In M. Takashi (Ed.), *Little boy: The arts of Japan's exploding subculture* (L. Hoaglund, Trans.) (pp. 187–207). New York/New Haven/London: Japan Society and Yale University Press.

Sawaragi, N. (2005b). *Sensō to banpaku* [War and expo]. Tokyo: Bijutsu Shuppansha.

Sawayaka. (2008). Kumiawasareru shōjo [A constructed girl]. *Yuriika, 40*(15), 184–192.

Sawayaka. (2014). *Jūnendai Bunkaron* [Cultural theory concerning 2010s]. Tokyo: Seikaisha Shinsho.

Schechner, R. (2013). *Performance studies : An introduction* (3rd ed.). New York: Routledge.

Scheer, E. (2015). Robotics as new media dramaturgy: The case of the sleepy robot. *TDR: The Drama Review, 59*(3), 140–149. doi:10.1162/DRAM_a_00476.

Schodt, F. L. (1988). *Inside the robot kingdom: Japan, mechatronics, and the coming robotopia.* Tokyo/New York: Kodansha International.

Searle, J. R. (1980). Minds, brains, and programs. *Behavioral and Brain Sciences, 3* (03), 417–424. doi:10.1017/S0140525X00005756.

Sena, H. (2001). *Robotto Seiki.* Tokyo: Bungei Shunjū.

Sena, H. (Ed.). (2004). *Robotto Opera (Robot Opera: An anthology of robot fiction and robot culture, original English title).* Tokyo: Kōbunsha.

Sena, H. (2008). *Sena Hideaki robottogaku ronshū* [Hideaki Sena robot study essay collection]. Tokyo: Keisō Shobō.

Sharkey, N., & Sharkey, A. (2012). The rights and wrongs of robot care. In L. Patrick, A. Keith, & G. A. Bekey (Eds.), *Robot ethics: The ethical and social implications of robotics* (pp. 267–282). Cambridge, MA: MIT Press.

Shaw-Garlock, Glenda. (2014). Gendered by design: Gender codes in social robotics. In *Sociable robots and the future of social relations: Proceedings of robophilosophy 2014* (pp. 309–317). Amsterdam: IOS Press.

Shershow, S. C. (1995). *Puppets and 'popular' culture*. Ithaca: Cornell University Press.

Shiba, T. (2014). *Hatsune miku wa naze sekai wo kaetanoka* [Why has hatsune Miku changed the world]. Tokyo: Ōta Shuppan.

Shibata, M. (2001). *Robotto no kokoro: Nanatsu no tetsugaku monogatari* [The robot's soul: Seven philosophical stories]. Tokyo: Kōdansha.

Shibata, T. (2010). Ippankatei ni okeru azarashigata robotto Paro tono fureai [Interaction with Paro at Home]. *Robotto, 197*(November), 24–27.

Shibata, T. (2012). Serapī yō robotto Paro no kenkyū kaihatsuto kokunaigai no dōkō [Local and international trends concerning the study and development of therapy robot Paro]. In *Azarashigata robotto Paro niyoru robotto serapī kenkyūkai, shōrokushū* [Therapy robot study group with seal-like Paro, short essay collection] (pp. 4–17). Tokyo: AIST, Shutodaigaku Tokyo, IEEE RAS.

Shibata, T. (2015). Azarashigata robotto niyoru shinkeigaku teki serapī [Neurological therapy with seal-like Paro]. *Seimitsukō Gakkaishi, 81*(1), 18–21.

Shibata, T., & Yukawa, T. (2014). Mentaru komittomento robotto Paro, Fukushi no genba deno katsuyō to iyashi [Mental commitment robot Paro: Use and therapy at a welfare facility]. *Gekkan Fukushi, 97*(8).

Shigematsu, S. (1999). Dimensions of desire: Sex, fantasy, and fetish in Japanese comics. In J. A. Lent (Ed.), *Themes and issues in Asian cartooning: Cute, cheap, mad, and sexy* (pp. 127–163). Bowling Green: Bowling Green State University Popular Press.

Silverberg, M. R. (2006). *Erotic grotesque nonsense the mass culture of Japanese modern times*. Berkeley: University of California Press.

Smith, M. (Ed.). (2005). *Stelarc: The monograph*. Cambridge, MA/London: MIT Press.

Sonoyama, T. (2007). *Robotto dezain gairon* [Introduction to robot design]. Tokyo: Mainichi Komyunikēshonzu.

Sorell, T., & Draper, H. (2014). Robot carers, ethics, and older people. *Ethics and Information Technology, 16*(3), 183–195. doi:10.1007/s10676-014-9344-7.

Sparrow, R., & Sparrow, L. (2006). In the hands of machines? The future of aged care. *Minds and Machines, 16*(2), 141–161. doi:10.1007/s11023-006-9030-6.

Stalker, D., & Glymour, C. (1982). The malignant object: Thoughts on public sculpture. *The Public Interest, 66*, 3–21.

Starrs, R. (2011). *Modernism and Japanese culture*. Houndmills/Basingstoke/New York: Palgrave Macmillan.

Steinberg, M. (2012). *Anime's media mix franchising toys and characters in Japan*. Minneapolis: University of Minnesota Press.

Stevens, C. S. (2010). You are what you buy: Postmodern consumption and fandom of Japanese popular culture. *Japanese Studies, 30*(2), 199–214. doi:10.1080/10371397.2010.497578.

Sugano, S. (2011). *Hito ga mita yume, robotto no kita michi: Girisha shinwa kara Atomu, soshite* [Mankind's Dream, The Historical Path: The Robot From (ancient) Greece to Astro Boy]. Tokyo: JIPM Sorūshon.

Sugimoto, Y., & Mouer, R. (1995). *Nihonjinron no hōteishiki* [The Equation of Nihonjinron]. Tokyo: Chikuma Shobō.

Suzuki, K. (Ed.). (2007a). *Dai robotto haku: Karakuri kara anime, saishin robotto made* [The Great Robot Exhibition: From Karakuri, to anime, to the latest robots]. Exh. cat. Tokyo: Yumiuri Shimbun.

Suzuki, K. (2007b). Karakuri ga hagukunda nihon no robotto kan. In K. Suzuki (Ed.), *Dai robotto haku: Karakuri kara anime, saishin robotto made* [The Great Robot Exhibition: From karakuri, to anime, to the latest robots] (pp. 44–49). Exh. cat. Tokyo: Yumiuri Shimbun.

Suzuki, H., Nishi, H., & Taki, K. (2009). Ningen no kansei ni motozuku dōbutsugata robotto no tameno yonkyaku hoyō seisei [Generation method of quadrupedal gait based on human feeling for animal type robot]. *Chinō to Jōhō, 21*(5), 653–662.

Suzumori, K. (2012). *Robotto wa naze ikimono ni niteshimaunoka: Kōgaku ni tachihadakaru kyūkyoku no rikigaku kōzō* [How would a robot become similar to a living creature: The ultimate principle of dynamics before engineering]. Tokyo: Kōdansha.

Tait, P. (2012). *Wild and dangerous performances: Animals, emotions, circus.* Houndmills/Basingstoke/Hampshire/New York: Palgrave Macmillan.

Takahashi, N. (2011). *Bōkaroido genshō: Sinsēki kontentsu sangyō no mirai moderu* [The phenomenon of Vocaloid: The future model for content business in the new century]. Tokyo: PHP Kenkyūjo.

Takahashi, H. (2016). KABUTOM RX-3: Beetle robot official HP. http://kabutom.com/. Accessed on 20-01-2016.

Takahashi, T., & Yanagida, R. (2011). *Sijō saikyō no robotto* [The strongest robot]. Tokyo: Media Fakutorī.

Takano, Y. (2008). *Shūdan shugi toiu sakkaku: Nihonjinron no omoichigai to sono yurai* ['Groupism' as Illusion: The Misunderstanding Concerning Nihonjinron and its Origin]. Tokyo: Shinchōsha.

Takatsuki, Y. (2009). *Rorikon: Nihon no shōjo shikōsha tachi to sono sekai* [Lolicon: Japan's Shōjo Lovers and Their World]. Tokyo: Basilico.

Takenishi, M. (Ed.). (2005). *Aichi banpaku saishin robotto gaido* [Aichi Expo. The guide to the latest robots]. Bessatsu Robocon Magazine. Tokyo: Ōmusha.

Takeno, J. (2011). *Kokoro wo motsu robotto: Hagane no shikō ga kagami no nakano jibunni kizuku* [The robot with soul: A realisation of the self in the mirror by a metallic thinking entity]. Tokyo: Nikkan Kōgyōsha.

Takeuchi, Y. (2007). Ējent medieiteddo intarakushon [Agent-mediated interaction]. In S. Yamada (Ed.), *Hito to robotto no aida wo dezainsuru* [Designing that which is in-between Humans and Robots] (pp. 259–288). Tokyo: Tokyo Denki Daigaku Shuppankyoku.

REFERENCES 249

Tamagawa, H. (2012). Comic market as space for self-expression in otaku culture. In M. Itō, D. Okabe, & I. Tsuji (Eds.), *Fandom unbound: Otaku culture in a connected world.* (pp. 107–132). New Haven: Yale University Press.

Tamura, T., Yonemitsu, S., Itoh, A., Oikawa, D., Kawakami, A., Higashi, Y., Fujimooto, T., & Nakajima, K. (2004). Is an entertainment robot useful in the care of elderly people with severe dementia? *The Journals of Gerontology. Series A, Biological Sciences and Medical Sciences, 59*(1), 83–85.

Tanaka, Y. (2010). War and Peace in the Art of Tezuka Osamu: The humanism of his epic manga. *The Asia-Pacific Journal: Japan Focus, 8*(38.1), 1–15.

Tanaka, F., Cicourel, A., & Movellan, J. R. (2007). Socialization between toddlers and robots at an early childhood education center. *Proceedings of the National Academy of Sciences of the United States of America, 104*(46), 17954–17958.

Tane, K. (2010). *Gandamu to nihonjin* [Gundam and the Japanese]. Tokyo: Bungē shunjū.

Taniguchi, T. (2010). *Komunikeishon suru robotto wa tsukureruka* [Is it possible to build a communication robot]. Tokyo: NTT shuppan.

Taylor, C. (2004). *Modern social imaginaries.* Durham/London: Duke University Press.

Tezuka, O. (1996). *Garasuno chikyū wo sukue: Nijū isseiki no kimitachi e* [Save the glass earth: For you, the 21st-century generation]. Tokyo: Kōbunsha.

The Humanoid Research Group of National Institute of Advanced Industrial Science and Technology. (2015). Hataraku ningengata robotto HRP-2 purototaipu wo kaihatsu [Developing working humanoid HRP-2, prototype]. https://unit.aist.go.jp/is/humanoid/m_projects/hrp-2p_j.html. Accessed on 09-15-2015.

The International Federation of Robotics (IFR). (2015). Industrial robots: IFR International Federation of Robotics. http://www.ifr.org/industrial-robots/. Accessed on 10-05-2015.

The Ministry of Economy, Trade and Industry. (2005). Robotto seisaku kenkyūkai chūkan hōkokusho [Research Group on Robot Policy Interim Report]. http://www.meti.go.jp/policy/robotto/chukanhoukoku.pdf. Accessed on 08-20-2015.

The Ministry of Economy, Trade and Industry. (2014, June 24). Japan revitalization strategy (Revised in 2014)-Japan's challenge for the future. https://www.kantei.go.jp/jp/singi/keizaisaisei/pdf/honbun2JP.pdf. Accessed on 12-20-2015.

The Ministry of Economy, Trade and Industry. (2015a, January 23). Japan's robot strategy was compiled- action plan toward a new industrial revolution driven by robots. http://www.meti.go.jp/english/press/2015/0123_01.html. Accessed on 06-15-2015.

The Ministry of Economy, Trade and Industry. (2015b). New robot strategy: Japan's robot strategy: Vision, strategy, action plan. http://www.meti.go.jp/english/press/2015/pdf/0123_01b.pdf. Accessed on 06-15-2015.

The Statistics Bureau, Ministry of Internal Affairs and Communications. (2013, September 15). Kōreisha no jinkō [The elderly population]. *Tōkei Topikkusu*

No. 72. http://www.stat.go.jp/data/topics/topi721.htm. Accessed on 05-20-2015.

The Statistics Bureau, Ministry of Internal Affairs and Communications. (2016a, February 26). Heisei 27 nen kokusei chōsa [The 2015 population census]. http://www.stat.go.jp/data/kokusei/2015/kekka.htm. Accessed on 03-15-2016.

The Statistics Bureau, Ministry of Internal Affairs and Communications. (2016b, July 29). Chūshutsu sokuhō shūkei kekka [The statistics result of the preliminary sample tabulation]. http://www.stat.go.jp/data/kokusei/2015/kekka.htm. Accessed on 06-30-2016.

The Yomiuri Shimbun. (2016, July 1). Gunji, Minsei Kakine Koeru Gijutsu: Bōeiyosan, Hantai No Daigaku Mo (Technologies that cross the wall between military and civilian use: A defense budget and opposing universities). Yomiuri Online. http://www.yomiuri.co.jp/osaka/feature/CO022791/20160701-OYTAT50015.html. Accessed on 07-30-2016.

Tillis, S. (1992). *Toward an aesthetics of the puppet: Puppetry as a theatrical art.* New York: Greenwood Press.

Toida, M. (1994). *Engi* [Acting]. Tokyo: Kinokuniya Shoten.

Toyota Motor Corporation. (2015). Toyota Global Site, Expo 2005 Aichi. http://www.toyota-global.com/innovation/partner_robot/aichi_expo_2005/index04.html. Accessed on 10-10-2015.

Tufnell, N. (2014, March 27). Cybathlon 2016: First 'olympics' for bionic athletes. *Wired UK.* http://www.wired.co.uk/news/archive/2014-03/27/cybathlon. Accessed on 01-20-2016.

Turkle, S. (2007). *Evocative objects things we think with.* Cambridge, MA: MIT Press.

Turkle, S. (2011). *Alone together: Why we expect more from technology and less from each other.* New York: Basic Books.

Turner, V. (1967). *The forest of symbols: Aspects of Ndembu ritual.* Ithaca: Cornell University Press.

Turner, V. (1982). *From ritual to theatre: The human seriousness of play.* New York: PAJ Publications.

Turner, V. (1988). *The anthropology of performance.* New York: PAJ Publications.

Ueno, T. (2001). Japanimation and techno-orientalism. In G. Bruce (Ed.), *The uncanny: Experiments in cyborg culture* (pp. 223–231). Vancouver: Vancouver Art Gallery and Arsenal Pulp Press.

Ueno, C. (2010). *Onnagirai: Nippon no misoginī* [Woman Hater: Japanese Misogyny]. Tokyo: Kinokuniya Shoten.

Umetani, Y. (2005). *Robotto no kenkyūsha wa gendai no karakurisi ka?* [Are robot researchers contemporary Karakuri Craftsmen?]. Tokyo: Ōmusha

Umezawa, M., & Nakamura, Y. (2012). Hatsune Miku genshō ga hiraku kyōkanryoku no shinsekai [The New World of Empathy Through the Hatsune

Miku Phenomenon]. Interview with Hiroyuki Ito, Yasuharu Ishikawa, and Toshiyuki Inoko. *Toyo Keizai On-Line*. http://toyokeizai.net/articles/-/11315. Accessed on 15-04-2013.

Uno, T. (2011a). *Ritoru pīpuru no jidai* [The age of little people]. Tokyo: Gentōsha.

Uno, T. (2011b). *Zeronendai no sōzōryoku* [The 'Noughties' imagination]. Tokyo: Hayakawa Shobō.

Uno, T. (Ed.). (2015). *Planets: Tokyo 2020 alternative Olympic project* (Vol. 9). Tokyo: Dainiji Wakusei Kaihatsu Iinkai/Planets.

Varney, D., Eckersall, P., Hudson, C., & Hatley, B. (2013). *Theatre and performance in the Asia-Pacific: Regional modernities in the global era*. Houndmills/Basingstoke/Hampshire/New York: Palgrave Macmillan.

Veltruský, J. (1964). Man and object in the theater. In P. L. Garvin (Ed.), *A Prague school reader on esthetics, literary structure, and style* (pp. 83–91). Washington, DC: Georgetown University Press.

Vincent, J. K. (2011). Translator's introduction, making it real: Fiction, desire, and the queerness of the beautiful fighting girl. In S. Tamaki (Ed.), *Beautiful fighting girl* (J. K. Vincent & D. Lawson, Trans.) (pp. ix–xxv). Minneapolis: University of Minnesota Press.

Wada, K. (2012). Kōreisha shisetsu ni okeru Paro no unyō hōhō [The method of use for Paro at aged care facilities]. In *Azarashigata robotto Paro ni yoru robotto serapī kenkyūkai, shōrokushū* [*Therapy robot study Group with seal-like Paro, short essay collection*] (pp. 18–20). Tokyo: AIST, Shutodaigaku Tokyo, IEEE RAS.

Wada, K., & Shibata, T. (2007). Living with seal robots: Its sociopsychological and physiological influences on the elderly at a care house. *IEEE Transactions on Robotics, 23*(5), 972–980. doi:10.1109/TRO.2007.906261.

Wallach, W., & Allen, C. (2009). *Moral machines: Teaching robots right from wrong*. Oxford/New York: Oxford University Press.

Walton, K. L. (1990). *Mimesis as make-believe: On the foundations of the representational arts*. Cambridge, MA: Harvard University Press.

Wang, N. (1999). Rethinking authenticity in tourism experience. *Annals of Tourism Research, 26*(2), 349–370. doi:10.1016/S0160-7383(98)00103-0.

Watanabe, T. (2003). Shintai komyunikēshon ni okeru hikikomi to shintaisei: Kokoro ga kayou shintaiteki komyunikēshon sisutemu E-Cosmic no kaihatsu wo tōshite [The phenomenon of entrainment and the body in bodily communication: The expression of soul in the development of communication system e-cosmic]. *Bebī Saiensu, 2*, 4–12.

Weiss, A., Wurhofer, D., & Tscheligi, M. (2009). "I love this dog"—Children's emotional attachment to the robotic dog AIBO. *International Journal of Social Robotics, 1*(3), 243–248. doi:10.1007/s12369-009-0024-4.

Weitz, E. (2016). *Theatre and laughter*. London/New York: Palgrave Macmillan.

Weston, H. (2009). Fables and follies: Florian's 'The monkey showing the magic lantern' and the failure of imitation. In A. Satz & J. Wood (Eds.), *Articulate objects: Voice, sculpture and performance* (pp. 47–62). Oxford/New York: Peter Lang.

White, G. (2013). *Audience participation in theatre: Aesthetics of the invitation.* Houndmills/Basingstoke/Hampshire/New York: Palgrave Macmillan.

Wiles, D. (2003). *A short history of Western performance space.* New York: Cambridge University Press.

Wilson, S. (2012). Exhibiting a new Japan: The Tokyo Olympics of 1964 and Expo '70 in Osaka. *Historical Research, 85*(227), 159–178. doi:10.1111/j.1468-2281.2010.00568.x.

Wood, G. (2003). *Edison's Eve: A magical history of the quest for mechanical life.* New York: Anchor Books.

Yablonsky, L. (1972). *Robopaths.* Indianapolis: Bobbs-Merrill.

Yamada, N. (2013). *Robotto to nihon: Kingendai bungaku, sengo manga niokeru jinkō shintai no hyōsō bunseki* [The robot and the human: An analysis of representations of the artificial body in modern literature and Postwar Manga]. Tokyo: Rikkyō Daigaku Shuppan.

Yamada, H. (2014, January 8). Jinkō chino gakkai no ayamari [The problem caused by the Japanese Society for Artificial Intelligence]. *The Huffington Post Japan.* http://www.huffingtonpost.jp/hajime-yamada/post_6588_b_4560115.html. Accessed on 10-10-2015.

Yamaguchi, M. (2002). The ludic relationship between man and machine in Tokugawa Japan. In J. Hendry & M. Raveri (Eds.), *Japan at play: The ludic and logic of power* (pp. 72–83). London/New York: Routledge.

Yamaguchi, T. (2006). Kateiyō robotto shijō wo kaitakushita AIBO [AIBO, which has opened the home robot market]. In P. H. P. Kenkyūjo (Ed.), *Otonano tameno robotto gaku* [Study of the robot for adults] (pp. 225–251). Tokyo: PHP Kenkyūjo.

Yamaguchi, M. (2007). *Dōke no minzokugaku.* Tokyo: Iwanami Shoten.

Yamanaka, H. (2006). Botomuzu wo tsukutte shimatta otoko, kataru, Part 1 [A Talk by the Man who Built Votoms, Part 1]. *NBOnline Premium.* http://business.nikkeibp.co.jp/free/x/20060328/20060328005467.shtml. Accessed on 10-01-2014.

Yamato, N. (2006). *Robotto to kurasu: Kateiyō robotto saizensen* [Living with a robot: The frontline of domestic robots]. Tokyo: Sofuto Banku Kurieitibu.

Yamatogokoro. (2013). Kabukichō no robotto resutoran ni naze gaikoku kyaku ga afureteirunoka? [Why does robot restaurant in Kabukicho attract so many foreign tourists?] *Yamatogokoro.jp.* http://www.yamatogokoro.jp/inbound-interview/index06.html. Accessed on 12-15-2013.

Yanobe, K. (2002). Ano Deme ga kaettekita [Deme Has Returned]. In Kirin Puraza Osaka (Ed.), *Expose 2002* (pp. 26–27). Osaka: Kirinbīru KPO Kirin Puraza Osaka.

Yanobe, K. (2005). Rocking Mammoth. *Kenji Yanobe Archive Project.* http://www.yanobe.com/artworks/rockingmammoth.html. Accessed on 05-01-2015.

Yano Research Institute. (2016). Kaigo robotto shijō ni kansuru chōsa wo jisshi (Implementation of research concerning the market for nursing care robots), Press Release. https://www.yano.co.jp/press/press.php/001546. Accessed on 10-10-2016.

Yokoyama, A. (2012). Rinshō seishinigaku kara mita, robotto serapī no mirai: Animaru serapī no chiken mo kangamite [The future of robot therapy from the perspective of clinical psychiatry: With knowledge of animal therapy]. *Keisokuto Seigyo, 51*(7), 598–602.

Yomota, I. (2006). *Kawaii ron* [A theory of cute]. Tokyo: Chikuma Shobō.

Yonemura, M. (2004). Atomu Ideorogī [Atom ideology]. In N. Baba (Ed.), *Robotto no bunkashi: Kikai wo meguru* sōzōryoku [Cultural analysis on the robot: The imagination concerning machines] (pp. 74–105). Tokyo: Shinwasha.

Yoneoka, T. (2012). Kōreisha shisetsu deno robotto serapī [Robot therapy at aged care facilities]. *Keisokuto Seigyo, 51*(7), 609–613.

Yonezawa, Y. (Ed.). (2002). *Robotto manga wa jitsugen suruka* [Could Robot Manga be Actualised]. Tokyo: Jitsugyō no Nihonsha.

Yoshida, M. (2004). *Nijigen bishōjo ron* [A theory of the two-dimensional beautiful girl]. Tokyo: Futami Shobō.

Yoshimi, S. (1992). *Hakurankai no seijigaku* [The Politics of Expo]. Tokyo: Chūkō Shinsho.

Yoshimi, S. (1999). "Made in Japan": The cultural politics of 'home electrification' in Postwar Japan. *Media, Culture & Society, 21*(2), 149–171. doi:10.1177/016344399021002002.

Yoshimi, S. (2005). *Banpaku gensō: Sengo seiji no jubaku* [Fantasies Through the Exposition: The Curse of Postwar Politics]. Tokyo: Chikuma Shinsho.

Yoshimi, S. (2008). *Toshi no doramatrugī: Tokyo sakariba no rekishi* [Dramaturgy of the city: History of entertainment districts in Tokyo]. Tokyo: Kawade shobō.

Yoshimi, S. (2015). Posuto sensō to shite no orinpikku: 1964 nen Tokyo taikai wo saikō suru [The Olympics as an alternative to war: A reexamination of the 1964 Tokyo Olympics]. *Masu Komyunikēshon Kenkyū, 86*(January), 19–37.

Yoshimoto, T. (2009). *Otaku no kigen* [The origins of the *otaku*]. Tokyo: NTT shuppan.

Yoshimoto Creative Agency. (2008). Bacarobo 2008. http://bacarobo.com. Accessed on 10-15-2009.

Zhao, S. (2006). Humanoid social robots as a medium of communication. *New Media & Society, 8*(3), 401–419. doi:10.1177/1461444806061951.

Zhu, Y. (2012). Performing heritage: Rethinking authenticity in tourism. *Annals of Tourism Research, 39*(3), 1495–1513. doi:10.1016/j.annals.2012.04.003.

Index

Note: Page numbers followed by "n" denote notes.

© The Author(s) 2017
Y. Sone, *Japanese Robot Culture*,
DOI 10.1057/978-1-137-52527-7

The manufacturer's authorised representative in the EU is Springer
Nature Customer Service Centre GmbH, Europaplatz 3, 69115 Heidelberg,
Germany. If you have any concerns regarding our products, please
contact ProductSafety@springernature.com

Printed and bound by CPI Group (UK) Ltd, Croydon, CR0 4YY
23/04/2026
02095587-0005